Nutrient Requirements and Fertilizer Management Strategies in Plant Cultivation

Nutrient Requirements and Fertilizer Management Strategies in Plant Cultivation

Guest Editor

Lin Tang

MDPI

Basel • Beijing • Wuhan • Barcelona • Belgrade • Novi Sad • Cluj • Manchester

Guest Editor
Lin Tang
Sustainable Field Crops Program
Institute of Agrifood Research and Technology (IRTA)
Lleida
Spain

Editorial Office
MDPI AG
Grosspeteranlage 5
4052 Basel, Switzerland

This is a reprint of the Special Issue, published open access by the journal *Plants* (ISSN 2223-7747), freely accessible at: https://www.mdpi.com/journal/plants/special issues/W3IZ8J0029.

For citation purposes, cite each article independently as indicated on the article page online and as indicated below:

Lastname, A.A.; Lastname, B.B. Article Title. *Journal Name* **Year**, *Volume Number*, Page Range.

ISBN 978-3-7258-7328-9 (Hbk)
ISBN 978-3-7258-7329-6 (PDF)
https://doi.org/10.3390/books978-3-7258-7329-6

Contents

About the Editor . . . vii

Preface . . . ix

Lin Tang
Nutrient Requirements and Fertilizer Management Strategies in Plant Cultivation
Reprinted from: *Plants* **2025**, *14*, 2981, https://doi.org/10.3390/plants14192981 . . . 1

Md. Jahidul Islam Shohag, Elena Máximo Salgado, Marina Curtis Gluck and Guodong Liu
Evaluating the Impact of Phosphorus and Solid Oxygen Fertilization on Snap Bean (*Phaseolus vulgaris* L.): A Two-Year Field Study
Reprinted from: *Plants* **2024**, *13*, 3384, https://doi.org/10.3390/plants13233384 . . . 5

Liping Cheng, Jiapan Lian, Xin Wang, Mehr Ahmed Mujtaba Munir, Xiwei Huang, Zhenli He, et al.
Evaluating a Soil Amendment for Cadmium Mitigation and Enhanced Nutritional Quality in Faba Bean Genotypes: Implications for Food Safety
Reprinted from: *Plants* **2025**, *14*, 141, https://doi.org/10.3390/plants14010141 . . . 25

Keru Yu, Zhenyu Wang, Wenyan Yang, Shuai Li, Dongtao Wu, Hongtao Zheng, et al.
Application of Biochar-Immobilized *Bacillus megaterium* for Enhancing Phosphorus Uptake and Growth in Rice
Reprinted from: *Plants* **2025**, *14*, 214, https://doi.org/10.3390/plants14020214 . . . 41

Davide Raffaelli, Rohullah Qaderi, Luca Mazzoni, Bruno Mezzetti and Franco Capocasa
Yield and Sensorial and Nutritional Quality of Strawberry (*Fragaria* × *ananassa* Duch.) Fruits from Plants Grown Under Different Amounts of Irrigation in Soilless Cultivation
Reprinted from: *Plants* **2025**, *14*, 286, https://doi.org/10.3390/plants14020286 . . . 59

Yanmin Peng, Kainan Zhao, Jun Zhang, Kaiming Ren, Junhao Zhang, Jinhua Guo, et al.
Subsoiling Before Wheat Sowing Enhances Grain Yield and Water Use Efficiency of Maize in Dryland Winter Wheat and Summer Maize Double Cropping System Under One-Off Irrigation Practice During the Wheat Season
Reprinted from: *Plants* **2025**, *14*, 738, https://doi.org/10.3390/plants14050738 . . . 74

Ignacio Darío Flores-Sánchez, Manuel Sandoval-Villa and Ebandro Uscanga-Mortera
Nutrient Uptake of Two Semidomesticated *Jaltomata* Schltdl. Species for Their Cultivation
Reprinted from: *Plants* **2025**, *14*, 1124, https://doi.org/10.3390/plants14071124 . . . 94

Yafei Wang, Juntao Liu, Yuelin He, Wei Zhu, Liming Jia and Benye Xi
Growth and Biomass Distribution Responses of *Populus tomentosa* to Long-Term Water–Nitrogen Coupling in the North China Plain
Reprinted from: *Plants* **2025**, *14*, 1833, https://doi.org/10.3390/plants14121833 . . . 113

Ying Wang, Fulai Yan, Junliang Fan and Fucang Zhang
Optimization of Irrigation Amount and Nitrogen Rate of Drip-Fertigated Sugar Beet Based on Sugar Yield, Nitrogen Use Efficiency, and Critical Nitrogen Dilution Curve in the Arid Southern Xinjiang of China
Reprinted from: *Plants* **2025**, *14*, 2055, https://doi.org/10.3390/plants14132055 . . . 132

About the Editor

Lin Tang

Lin Tang is a researcher participating in the Sustainable Field Crops Program at the Institute of Agrifood Research and Technology (IRTA), Spain. Her research focuses on evaluating ecosystem services to promote legume cultivation across Europe. Her primary interest lies in leveraging plant–soil–microbe interactions to develop cropping systems with high protein quality and low nutrient emissions, while investigating the underlying mechanisms influencing improved crop performance, nutrient cycling and overall system resilience. She also has extensive experience in soil organic carbon sequestration, greenhouse gas mitigation, forage production, phytoremediation and biofortification.

Preface

This Reprint, entitled "Nutrient Requirements and Fertilizer Management Strategies in Plant Cultivation", compiles a selection of peer-reviewed contributions addressing the scientific and practical challenges of plant nutrient management. Adequate and balanced nutrient supply is fundamental to plant growth, reproductive development and yield formation, while nutrient deficiencies or imbalances may compromise productivity and environmental sustainability. As agricultural systems face increasing pressure to enhance food production while reducing ecological impacts, improving nutrient use efficiency has become a critical objective.

The studies assembled in this Reprint examine nutrient requirements and fertilizer management from agronomic, physiological and ecological perspectives, providing an integrated view across varied crop species and cultivation systems. The contributions highlight not only the direct effects of fertilization practices on nutrient uptake and utilization but also the interactions between nutrient management, water regulation and soil improvement that influence nutrient dynamics and plant performance.

By integrating fundamental research with applied agronomic practices, this Reprint will advance understanding of nutrient dynamics in cultivation systems and support the development of efficient, sustainable and environmentally responsible fertilization strategies. It is intended to be read by researchers, agronomists and professionals engaged in crop production and sustainable agricultural management.

Lin Tang
Guest Editor

plants

Editorial

Nutrient Requirements and Fertilizer Management Strategies in Plant Cultivation

Lin Tang

Department of Agroecology, Aarhus University, 8830 Tjele, Denmark; lin.tang@agro.au.dk; Tel.: +45-50108877

1. Introduction

Essential plant nutrients are fundamental to plant growth and reproduction, and deficiencies in any of them can disrupt or even halt the plant growth cycle [1]. In cultivated systems, it is often necessary to supplement soil fertility through the judicious application of fertilizers in order to meet crop nutrient demands and maintain or increase yields [2,3]. Accurate nutrient diagnosis is a key step in determining plant nutritional status and provides crucial guidance for allocating fertilizers to producers and advisors [4,5].

Balanced fertilization refers to supplying essential nutrients in optimal quantities and appropriate proportions, using suitable application methods and timing to match the specific requirements of crop needs and agroclimatic conditions [6,7]. Developing novel and sophisticated fertilization strategies remains a major challenge in nutrient management [8,9]. Optimized nutrient management not only prevents deficiencies, imbalances or excessive fertilizer use, but also enhances nutrient-use efficiency, supports productive agronomic systems and promotes environmentally sustainable plant cultivation [10,11].

In such a context, this Special Issue (SI) aims at advancing our understanding of nutrient requirements and fertilizer management strategies in plant cultivation. It considers not only the direct impacts of fertilizer practices on nutrient uptake and utilization, but also the indirect effects of water management and soil improvement on nutrient dynamics. The SI comprises eight original research articles contributed by authors from institutions in China, Italy, Mexico and the United States. Together, these studies provide an integrated perspective on a wide range of diverse plant species, including legumes, cereals, fruits, medicinal herbs, trees and sugar crops cultivated in diverse systems such as hydroponics, substrate culture, pot experiments and field trials, and examined across physiological, molecular and ecological scales under different climatic conditions.

2. Advances in Nutrient Requirements and Fertilizer Management Strategies in Plant Cultivation

The following section presents an overview of the articles included in this SI, highlighting how their findings collectively advance research on nutrient requirements and fertilizer management strategies in plant cultivation.

Shohag et al. [12] conducted a two-year field experiment to evaluate the combined effects of phosphorus (P) and oxygen (O_2) fertilization on plant growth, pod yield and P uptake of snap bean (*Phaseolus vulgaris* L.) grown in acidic sandy soil. Their study showed that applying triple superphosphate together with calcium peroxide significantly enhanced vegetative growth and pod production. Additionally, strong associations were observed between growth, pod yield and nutrient accumulation, and notable seasonal differences emerged in both crop performance and soil P dynamics. This research highlights that incorporating O_2 fertilizers is a cost-effective strategy to alleviate hypoxic stress, improve P use efficiency and increase yields in snap bean cultivation.

Cheng et al. [13] investigated the effects of a soil amendment (SA) on agronomic performance, seed nutritional quality and cadmium (Cd) accumulation in several faba bean (*Vicia faba* L.) genotypes cultivated in acidic soils. Their study showed that SA application raised soil pH, lowered phytoavailable Cd and reduced Cd concentrations in seeds while alleviating soil manganese and aluminum toxicity. In addition, SA enhanced the levels of essential macronutrients and micronutrients in seeds, while reducing phytate content, with only a marginal decrease in average yield. This research provides valuable evidence of the potential of SA to improve nutrient fortification and mitigate Cd contamination in farmland affected by Cd pollution.

Yu et al. [14] conducted a series of pot experiments to exam how soil microbial P mobilization and rice P uptake respond to applications of rice husk biochar immobilized with *Bacillus megaterium* (BMB). Their results revealed that the treatment combining organic fertilizer with BMB (MOF) produced the most pronounced improvements in soil available P and fostered beneficial shifts in bacterial community structure, thereby enlarging the pool of bioavailable P reservoir. Moreover, the MOF treatment significantly enhanced soil carbon utilization and rice agronomic performance. This research offers a valuable theoretical basis for understanding the role of biochar-based bacterial inoculants in improving P management in rice systems.

Raffaelli et al. [15] conducted a two-year soilless greenhouse experiment to identify strawberry (*Fragaria* × *ananassa*) genotypes with superior capacity to sustain yield and enhance fruit quality under reduced irrigation. Their findings showed that moderate water restriction improved several quality attributes, including citric acid concentration, soluble solids content and color brightness, while fruit firmness remained unaffected. Total phenolic content exhibited the strongest response to water stress, indicating enhanced antioxidant activity. This research underscores the importance of selecting genotypes best suited to water regimes in order to enhance fruit quality, maintain high production standards and promote water-use efficiency in strawberry cultivation.

Peng et al. [16] conducted a three-year field experiment to quantify the effects of tillage methods prior to wheat sowing and irrigation practices during the wheat season on the yield formation and water-use efficiency of subsequent summer maize production. Their results showed that, compared with rotary tillage and conventional plowing, subsoiling before wheat sowing significantly increased soil water storage at maize sowing, thereby enhancing dry matter accumulation and nutrient uptake in maize. One-off irrigation during the wheat season negatively affected pre-sowing soil water storage and maize productivity; subsoiling effectively mitigates these adverse effects. Overall, tillage practices before wheat sowing exerted a greater influence on soil water storage and maize quality than irrigation strategies applied during the wheat growing season. This research provides practical guidance for optimizing soil water use and improving maize productivity in dryland systems.

Flores-Sánchez et al. [17] investigated nutrient uptake dynamics in two *Jaltomata* species grown under three levels of electrical conductivity in greenhouse hydroponic systems. Their study revealed marked variability in nutrient concentration and total dry matter (TDM) accumulation between species and across phenological stages. For macronutrients, *J. procubens* showed the highest P concentrations from the vegetative stage, magnesium up to fructification, potassium during the vegetative phase and nitrogen (N) at flowing, whereas *J. tlaxcala* accumulated more calcium during fructification. Regarding micronutrients, *J. thaxcala* exhibited greater iron uptake from the vegetative stage, higher boron, zinc and TDM during flowering, and manganese at fructification. This research suggests that tailored fertilization programs are necessary for each species to optimize nutrient supply and productivity.

Wang et al. [18] carried out a six-year field study to evaluate the long-term effects of drip irrigation and N application on growth, biomass allocation and the efficiency of water and fertilizer use in short-rotation triploid *Populus tomentosa* plantations. Their results indicated that the impacts of water and N treatments on the annual increment of diameter at breast height, tree height, stand basal area (BA_S), stand volume (V_S) and annual forest productivity (AFP) were strongly influenced by stand age. Moreover, irrigation significantly increased BA_S, V_S, AFP and total forest biomass of six-year-old stands, with irrigation exerting a more consistent effect than N fertilization. This research provides valuable guidance for optimizing water and fertilizer management in short-rotation poplar plantations.

Wang et al. [19] conducted a three-year field experiment to assess the effects of irrigation levels and N application rates on sugar yield, N use efficiency and soil nitrate-N (NO_3^--N) residues in drip-fertigated sugar beet cultivated in an arid region of China. Their results showed that sugar yield and N uptake increased and then plateaued with high N rate, whereas N use efficiency decreased. Greater irrigation volumes reduced residual NO_3^--N in soil profiles. Based on these findings, the authors developed a reliable N nutrition index for sugar yield, derived from a critical N dilution curve established from sugar beet dry matter. This research provides an important basis for improving water and N management in sugar beet production across arid and semi-arid regions worldwide.

3. Conclusions

In summary, this SI highlights the rapidly advancing scientific frontiers of nutrient requirements and fertilizer management strategies in plant cultivation. It brings together current research findings and provides valuable theoretical foundations as well as practical pathways for addressing the increasingly complex challenges in this field. As the Guest Editor, I am deeply grateful to all the authors for contributing their high-quality research to this SI and to the reviewers for their careful assessments, which ensured the scientific rigor of the published papers. I hope that the articles presented here will serve as a rich source of knowledge, inspiration and fresh perspectives on nutrient requirements and fertilizer management strategies in plant cultivation for both researchers and practitioners.

Conflicts of Interest: The author declares no conflict of interest.

References

1. Aye, H.N.; Masih, S. Role of Nutrients in Plants, Its Deficiency and Management. *Int. J. Plant Soil Sci.* **2023**, *35*, 129–136. [CrossRef]
2. Wato, T.; Negash, T.; Andualem, A.; Bitew, A. Significance of Organic and Inorganic Fertilizers in Maintaining Soil Fertility and Increasing Crop Productivity in Ethiopia: A Review. *Environ. Res. Commun.* **2024**, *6*, 102002. [CrossRef]
3. Tang, L. Soil Fertility, Plant Nutrition and Nutrient Management. *Plants* **2025**, *14*, 34. [CrossRef] [PubMed]
4. Manzoor, R.; Akhtar, M.S.; Khan, K.S.; Raza, T.; Rehmani, M.I.A.; Rosen, C.; Rehman, M.K.U.; Zidan, N.; Alzuaibr, F.M.; Abdulsalam, N.M.; et al. Diagnosis and Recommendation Integrated System Assessment of the Nutrients Limiting and Nutritional Status of Tomato. *Phyton-Int. J. Exp. Bot.* **2022**, *91*, 2759–2774. [CrossRef]
5. Mamassi, A.; Tychon, B.; Hamma, A.; Balaghi, R.; El Gharous, M.; Marrou, H. Design and Evaluation of Fertilization Rule-Based Strategy with APSIM-Wheat Model: For a Climate Responsive and Risk Proofing Fertilization in Morocco. *Int. J. Plant Prod.* **2025**, *19*, 325–340. [CrossRef]

6. Tang, L.; Hamid, Y.; Liu, D.; Shohag, M.J.I.; Zehra, A.; He, Z.L.; Feng, Y.; Yang, X.E. Foliar Application of Zinc and Selenium Alleviates Cadmium and Lead Toxicity of Water Spinach–Bioavailability/Cytotoxicity Study with Human Cell Lines. *Environ. Int.* **2020**, *145*, 106122. [CrossRef] [PubMed]
7. Singh, A.; Khare, S.; Niharika; Gupta, P. Sulfur and Phosphorus Rransporters in Plants: Integrating Mechanisms for Optimized Nutrient Supply. *Plant Physiol. Bioch.* **2025**, *224*, 109918. [CrossRef] [PubMed]
8. Lian, J.P.; Cheng, L.P.; Zhai, X.; Wu, R.F.; Huang, X.W.; Chen, D.; Pan, J.Q.; Shohag, M.J.I.; Xin, X.P.; Ren, X.W.; et al. Zinc Glycerolate (Glyzinc): A Novel Foliar Fertilizer for Zinc Biofortification and Cadmium Reduction in Wheat (*Triticum aestivum* L.). *Food Chem.* **2023**, *402*, 134290. [CrossRef] [PubMed]
9. Mi, W.H.; Hong, Y.; Gao, F.; Ma, Y.Y.; Sun, T.; Wu, L.H.; Wang, G.Z.; Chen, S.T. Effect of Different Form of N Fertilization on Yield Sustainability and Soil Quality in Double Cropped Rice System in a Long-Term Experiment. *Soil Sci. Plant Nutr.* **2024**, *24*, 2815–2824. [CrossRef]
10. Quddus, M.A.; Ahmed, R.; Islam, M.; Haque, M.E.; Islam, M.A.; Alam, A.; Rahman, M.Z.; Fahad, Z.H.; Islam, M.K.; Gaber, A.; et al. Organic and Inorganic Fertilizers Influence the Productivity, Fruit Quality and Nutrient Use Efficiency of Strawberry (*Fragaria* × *ananassa* Duch.). *Sci. Rep.* **2025**, *15*, 26252. [CrossRef] [PubMed]
11. Tang, L.; Kumar, U.; Öhlund, L.; Parsons, D. Establishment and production of lucerne (*Medicago sativa* L.) in Sweden is affected by inoculation product choice. *J. Agric. Food Res.* **2025**, *19*, 101644. [CrossRef]
12. Shohag, M.J.I.; Salgado, E.M.; Gluck, M.C.; Liu, G.D. Evaluating the Impact of Phosphorus and Solid Oxygen Fertilization on Snap Bean (*Phaseolus vulgaris* L.): A Two-Year Field Study. *Plants* **2024**, *13*, 3384. [CrossRef] [PubMed]
13. Cheng, L.P.; Lian, J.P.; Wang, X.; Munir, M.A.M.; Huang, X.W.; He, Z.L.; Xu, C.J.; Tong, W.B.; Yang, X.E. Evaluating a Soil Amendment for Cadmium Mitigation and Enhanced Nutritional Quality in Faba Bean Genotypes: Implications for Food Safety. *Plants* **2025**, *14*, 141. [CrossRef] [PubMed]
14. Yu, K.R.; Wang, Z.Y.; Yang, W.Y.; Li, S.; Wu, D.T.; Zheng, H.T.; Ye, Z.Q.; Yang, S.N.; Liu, D. Application of Biochar-Immobilized *Bacillus megaterium* for Enhancing Phosphorus Uptake and Growth in Rice. *Plants* **2025**, *14*, 214. [CrossRef] [PubMed]
15. Raffaelli, D.; Qaderi, R.; Mazzoni, L.; Mezzetti, B.; Capocasa, F. Yield and Sensorial and Nutritional Quality of Strawberry (*Fragaria* × *ananassa* Duch.) Fruits from Plants Grown Under Different Amounts of Irrigation in Soilless Cultivation. *Plants* **2025**, *14*, 286. [CrossRef] [PubMed]
16. Peng, Y.M.; Zhao, K.N.; Zhang, J.; Ren, K.M.; Zhang, J.H.; Guo, J.H.; Wang, R.R.; Xiao, H.S.; Jiang, P.P.; Xu, N.L.; et al. Subsoiling Before Wheat Sowing Enhances Grain Yield and Water Use Efficiency of Maize in Dryland Winter Wheat and Summer Maize Double Cropping System Under One-Off Irrigation Practice During the Wheat Season. *Plants* **2025**, *14*, 738. [CrossRef] [PubMed]
17. Flores-Sánchez, I.D.; Sandoval-Villa, M.; Uscanga-Mortera, E. Nutrient Uptake of Two Semidomesticated *Jaltomata* Schltdl. Species for Their Cultivation. *Plants* **2025**, *14*, 1124. [CrossRef] [PubMed]
18. Wang, Y.F.; Liu, J.T.; He, Y.L.; Zhu, W.; Jia, L.M.; Xi, B.Y. Growth and Biomass Distribution Responses of *Populus tomentosa* to Long-Term Water-Nitrogen Coupling in the North China Plain. *Plants* **2025**, *14*, 1833. [CrossRef] [PubMed]
19. Wang, Y.; Yan, F.L.; Fan, J.L.; Zhang, F.C. Optimization of Irrigation Amount and Nitrogen Rate of Drip-Fertigated Sugar Beet Based on Sugar Yield, Nitrogen Use Efficiency, and Critical Nitrogen Dilution Curve in the Arid Southern Xinjiang of China. *Plants* **2025**, *14*, 2055. [CrossRef] [PubMed]

plants

MDPI

Article

Evaluating the Impact of Phosphorus and Solid Oxygen Fertilization on Snap Bean (*Phaseolus vulgaris* L.): A Two-Year Field Study

Md. Jahidul Islam Shohag, Elena Máximo Salgado, Marina Curtis Gluck and Guodong Liu *

Horticultural Sciences Department, Institute of Food and Agricultural Sciences, University of Florida, Gainesville, FL 32611, USA; jshohag@ufl.edu (M.J.I.S.); emaximosalgado@ufl.edu (E.M.S.); m.curtis@ufl.edu (M.C.G.)
* Correspondence: guodong@ufl.edu; Tel.: +1-352-273-4814; Fax: +1-352-846-0909

Abstract: The snap bean (*Phaseolus vulgaris* L.) is highly sensitive to both phosphorus (P) deficiency and hypoxic stress, which together can significantly hinder plant growth, nutrient uptake, and yield; however, limited information exists on the effect of P and oxygen (O_2) fertilization to alleviate these stresses and enhance yield. A two-year field experiment assessed the effects of P and O_2 fertilization on plant growth, pod yield, and P uptake in acidic sandy soil. Using a randomized complete block design with four replications, we tested five P rates (0, 45, 90, 135, and 179 kg ha^{-1} of phosphorus pentoxide, P_2O_5) in the form of triple superphosphate (TSP) along with two rates (0 and 45 kg ha^{-1}) of solid O_2 fertilizer as calcium peroxide (CaO_2). Phosphorus and O_2 fertilizers improved plant growth and pod yield, with the highest yield from the combination of 135 kg ha^{-1} P_2O_5 and 45 kg ha^{-1} CaO_2. Pearson correlation analysis indicated strong associations between plant growth, pod yield, and nutrient accumulation. Principal component analysis (PCA) highlighted notable seasonal differences in snap bean and soil characteristics. This study provides essential insights into the use of O_2 fertilizers as a cost-effective approach to mitigate hypoxia, enhance P use efficiency, and improve yield in snap bean. Our findings may inspire the development of sustainable nutrient protocols for high-quality snap bean production and serve as a foundation for similar applications in other crops.

Keywords: oxygen fertilizer; green bean; phosphorus; plant growth; vegetable production

1. Introduction

The snap bean (*Phaseolus vulgaris* L.), commonly referred to as green bean, is a vital legume crop cultivated extensively for its nutritious pods. Differently from dry common beans, snap beans are grown both for fresh vegetables and for processing (canning or freezing). A significant source of protein, dietary fiber, vitamins, minerals, and soluble sugar, snap beans are a common vegetable in human diets worldwide [1]. Snap beans are also a good source of antioxidants, including flavonoids and carotenoids, which contribute to their health benefits, such as reducing inflammation and supporting heart health [2,3]. Snap beans are highly recommended for low-carbohydrate diets, and their low glycemic index (GI) makes them an excellent choice for individuals managing diabetes [4]. According to recent data from the Food and Agriculture Organization of the United Nations, the global annual production of snap beans is approximately 23 million Mg [5]. The United States is one of the major producers of processed snap beans, with states like Wisconsin being key contributors. In 2023, Wisconsin alone produced about 6.75 million hundredweight (CWT) (343,000 Mg) of snap beans, accounting for nearly 48% of the U.S. total production [6]. The global demand for snap beans has been steadily increasing, due to their health benefits and culinary versatility. However, the productivity and quality of snap bean crops are heavily influenced by soil fertility and nutrient management practices, with the availability of essential nutrients like nitrogen (N), phosphorus (P), and potassium (K) playing a critical role [7].

Phosphorus (P) is a macronutrient essential for plants, with limited availability and low mobility in the soil, and it plays a vital role in plant biochemical processes [8], such as photosynthesis [9,10] and carbohydrate and protein metabolism [11]. This limitation of P availability is a significant factor, constraining up to 67% of primary agronomic productivity in croplands [12]. To mitigate these limitations, excessive P fertilization is a common practice in agricultural systems worldwide. This practice aims to maximize crop yields; however, this has led to a substantial increase in global P fertilization rates, which rose from approximately 5 Tg Year^{-1} in 1961 to 18 Tg Year^{-1} in 2013, surpassing the estimated planetary boundary of 6–12 Tg Year^{-1} [13]. Despite this increased application, crops only utilize about 20% of the applied P fertilizer [14]. MacDonald et al. (2011) reported that 30–50% of the P fertilizer applied to U.S. croplands does not contribute to crop growth and instead accumulates in the soil. This issue is particularly pronounced in northeast Florida, where excessive P accumulation has occurred due to continuous crop cultivation since the 1890s and the application of chemical P fertilizers since the 1920s [15]. As P is a nonrenewable resource, previous forecasts have indicated that P reserves could be depleted within fifty to one hundred years [16]. Excessive P application not only threatens environmental pollution, global food security, and crop quality due to the antagonistic effects of P on micronutrient availability, such as Fe [17] and Zn [18], but also contributes to severe water eutrophication through surface runoff [19] and leaching into groundwater [20]. Therefore, it is imperative to explore and implement optimal P management strategies that support sustainable production within intensive farming systems.

Phosphorus deficiency in snap beans can lead to stunted growth, delayed maturity, reduced biomass production, and ultimately lower yields. Symptoms of P deficiency include dark green leaves, purpling of leaf margins, and poor root development [21,22]. To address P deficiency, farmers typically apply P fertilizers; however, its efficiency is influenced by several factors, including soil pH, texture, organic matter content, and the presence of other nutrients [21,23]. Oxygen (O_2) is another critical factor influencing plant growth, particularly root respiration. Adequate O_2 supply in the root zone is essential for aerobic respiration, which provides the energy needed for nutrient uptake and root growth [24,25]. In poorly drained soils or under waterlogging conditions, O_2 availability can become limited, leading to anaerobic conditions that inhibit root function and reduce nutrient uptake efficiency [26]. Such O_2 stress can complexly exacerbate the effects of nutrient deficiencies, while adequate O_2 levels can enhance the efficiency of nutrient use [27]. Calcium peroxide (CaO_2) is an environmentally friendly material that releases O_2 in the presence of water and is widely used to improve O_2 availability in agricultural systems. In crop cultivation, CaO_2 fertilizer significantly increased seed germination and seedling growth of direct-seeded rice under waterlogging conditions [28]. This effect is attributed to CaO_2 reducing the accumulation of fermentation products in roots and alleviating the peroxidation of leaves, thereby promoting root vigor and improving agronomic traits such as branches, plant height, and stem diameter, leading to accelerated dry matter accumulation and nutrient uptake [29]. Their findings suggest that ensuring sufficient O_2 supply through CaO_2 application may not only mitigate waterlogging stress but also enhance nutrient uptake, including phosphorus, thereby improving plant growth and yield potential. In the context of P fertilization, improving O_2 supply to the root zone through materials like CaO_2 may enhance P uptake and utilization, potentially leading to better plant growth and higher yields in P-deficient sandy soils.

Research on P fertilization in snap beans has demonstrated the positive effects of adequate P supply on plant growth, yield, and quality. Studies have shown that P fertilization can enhance root development, increase biomass production, and improve pod yield and quality [30]. However, the response of snap beans to P fertilization has been shown to vary depending on soil properties, environmental conditions, and management practices [31,32]. While the benefits of P fertilization are well-documented, the role of O_2 in enhancing P uptake and utilization has received less attention. Our previous studies have suggested that supplemental O_2 can improve root health and function, leading to better nutrient uptake and plant growth [33,34]. However, there is limited research on the combined effects of P and O_2 fertilization in snap beans, particularly under field conditions. Therefore, this study aimed to evaluate the response of snap beans to different doses of P and O_2 fertilization over two growing seasons in fall 2022 and fall 2023. The objectives were to assess the impact of these fertilizers on plant growth, biomass, yield, and nutrient uptake, particularly P, and analyze their effect on soil P dynamics and crop performance.

2. Results

2.1. Soil and Weather Condition in the Experimental Site

At the experimental site, Hastings, Florida, the predominant soil orders, such as Spodosols, often exhibit aquic moisture regimes due to poor aeration and periodic water saturation. These conditions can result in redoximorphic features, commonly observed as redox depletions or stagnic properties in the soil profile. The depths at which these occur may vary but are typically influenced by fluctuating water tables and soil drainage characteristics. For Spodosols and related orders in Florida, such conditions are frequently noted within the upper soil horizons, particularly where organic matter accumulation and sandy textures dominate [35]. Soil chemical properties were analyzed before planting in both years. Notable differences were found in soil available P concentrations (Table 1).

Table 1. Chemical properties and critical limit of the experimental soil (0–20 cm depth) before planting in 2022 and 2023.

Properties (Unit)	Value		
	2022	2023	[3] Critical Limit
[1] pH (H_2O at 25 °C)	5.9	5.8	-
CEC (meq/100 g)	6.5	6.4	-
Soil Organic Matter (%)	0.30	0.29	Low
[2] Available P (mg kg^{-1})	65.17	70.14	[4] High
[2] Available K (mg kg^{-1})	47.5	45.79	Medium
[2] Available Ca (mg kg^{-1})	424	475.69	Medium
[2] Available Mg (mg kg^{-1})	77	75.5	Medium
[2] Available S (mg kg^{-1})	16.50	14.75	Low
[2] Available Mn (mg kg^{-1})	6.5	6.7	Medium

[1] The soil pH starts at 5.8 and may decrease by one unit during the growing season due to nitrification and cation nutrient uptake. [2] Mineral nutrients are Mehlich-III extractable. [3] The critical limits (low, medium, high) are adapted for local soil nutrient standards as outlined by UF/IFAS. [4] Although soil phosphorus levels are considered high according to Mehlich-III and UF/IFAS guidelines, most of this extractable P may not be bioavailable to the plants due to soil fixation by high concentrations of aluminum and iron at the experimental site.

In fall 2022, soil available P concentrations were 65.17 mg kg^{-1}, whereas in fall 2023, these values were 70.14, mg kg^{-1}. Average air temperatures in 2022 ranged from to 11 °C to 26 °C (Figure 1A), while in 2023 they ranged from 8 °C to 27 °C (Figure 1B). Cumulative rainfall during the growing seasons of 2022 and 2023 was 75.44 mm and 486.41 mm, respectively (Figure 1A,B, right axis). The greatest rainfall recorded was 173.74 mm on 29 September 2022 and 27.94 mm on 16 November 2023. No leaching rain events as defined by UF/IFAS guidelines occurred during the growing seasons.

2.2. Crop Growth and Biomass Production

The line graphs illustrate dry biomass production over the growth periods of fall 2022 and fall 2023 with different P_2O_5 rates, with or without O_2 fertilizer (Figure 2A,B). In fall 2022 (Figure 2A), biomass production increased steadily across all treatments, with the control showing the lowest biomass production throughout the growth period. The application of 135 kg ha^{-1} P_2O_5 with O_2 fertilizer resulted in the greatest biomass production, reaching approximately 29.9 g $plant^{-1}$ by 56 DAS. The treatments with O_2 fertilizer consistently outperformed those without it, indicating a positive effect of O_2 fertilizer on biomass production. Similarly, in fall 2023 (Figure 2B), biomass production followed a similar trend, with the control again showing the lowest biomass accumulation. The combination of 135 kg ha^{-1} P_2O_5 and O_2 fertilizer yielded the greatest biomass, closely followed by the 179 kg ha^{-1} P_2O_5 with O_2 fertilizer treatment, both maximum near 32 g plant^{-1} by 56 DAS. The data show that higher P_2O_5 rates and the addition of O_2 fertilizer enhanced biomass production compared to lower rates and treatments without O_2 fertilizer.

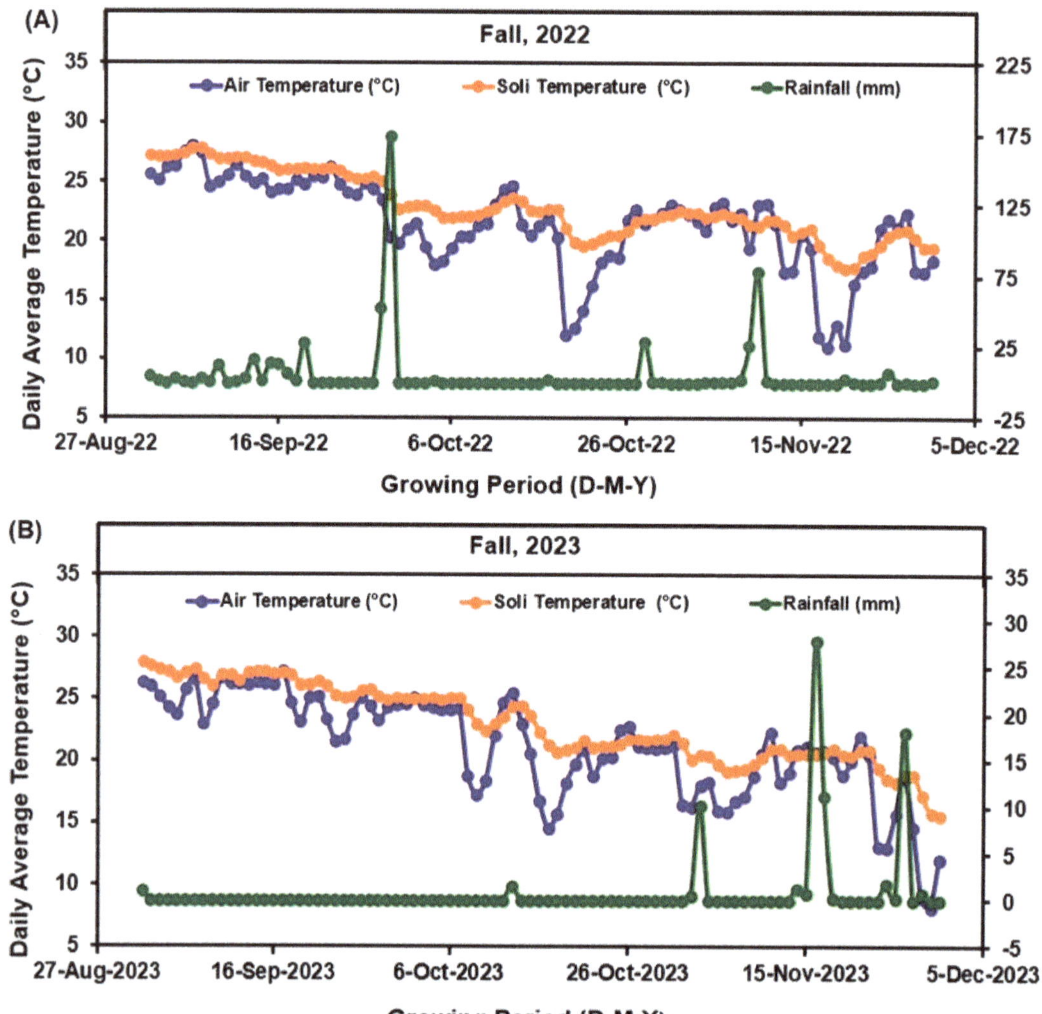

Figure 1. Daily average air temperature (°C), daily average soil temperature (°C), and daily average rainfall (mm) during the growing season of 2022 (**A**) and 2023 (**B**) from planting to harvesting date.

The boxplots in Figure 3A–D illustrate the effects of different P_2O_5 rates, with or without O_2 fertilizer, on SPAD readings and petiole NO_3^--N concentrations for the growing seasons of fall 2022 and fall 2023. For fall 2022 (Figure 3A), the control treatment had the lowest SPAD readings, indicating lower chlorophyll content. The SPAD readings increased with higher P_2O_5 rates, with 135 kg ha^{-1} P_2O_5 plus O_2 fertilizer showing the greatest SPAD values. Similarly, in fall 2023 (Figure 3B), the control treatment again exhibited the lowest SPAD readings. The application of 135 kg ha^{-1} P_2O_5 with O_2 fertilizer resulted in the greatest SPAD values, demonstrating the beneficial effect of combined P_2O_5 and O_2 fertilizer on chlorophyll content. In fall 2022 (Figure 3C), petiole NO_3^--N concentrations showed an increasing trend with higher P_2O_5 rates, with the greatest concentrations observed at 135 kg ha^{-1} P_2O_5 without O_2 fertilizer. However, treatments with O_2 fertilizer also demonstrated higher NO_3^--N levels compared to the control. In fall 2023 (Figure 3D), a similar trend was observed, with petiole NO_3^--N concentrations increasing with higher P_2O_5 rates. The greatest NO_3^--N concentrations were observed at 135 kg ha^{-1} P_2O_5 with O_2 fertilizer, followed closely by the 179 kg ha^{-1} P_2O_5 treatments. Overall, these results highlight the synergistic effect of P_2O_5 and O_2 fertilizer on promoting crop growth and biomass accumulation during both growing seasons.

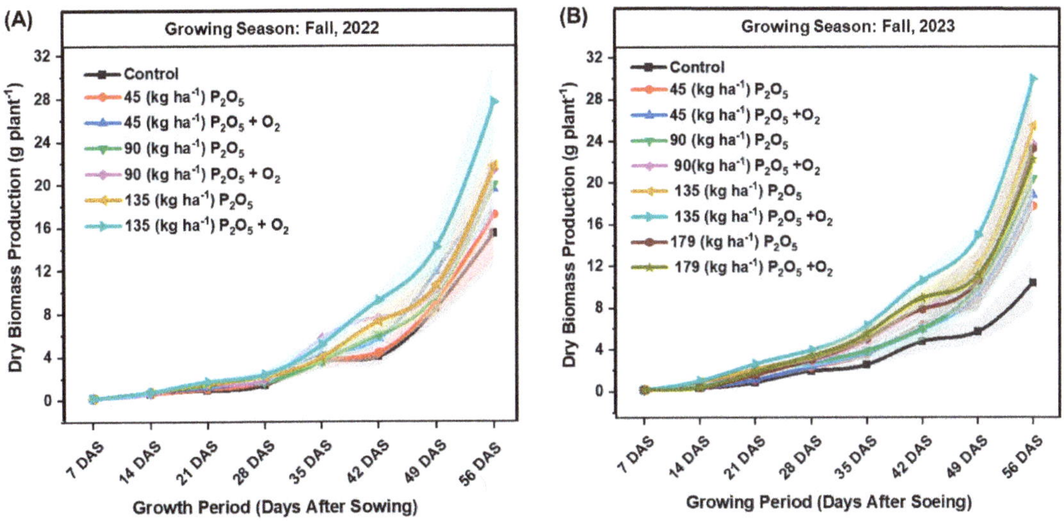

Figure 2. Effect of oxygen (O_2) fertilizer on dry biomass production for green beans grown under different rates of phosphorus (P_2O_5) fertilization in different growth stages during fall 2022 (**A**) and 2023 (**B**) at Hastings.

2.3. Pod Yield and Yield Attributes

The boxplot of pod yield in response to P_2O_5 rates, without or with O_2 fertilizer, demonstrated a wide variation among the treatments and across the years (Figure 4A,B). For fall 2022 (Figure 4A), the control treatment had the lowest pod yield, ranging between 5.1 and 6.7 Mg ha^{-1}. As the P_2O_5 rates increased, there was a notable rise in pod yield. At 45 kg ha^{-1} P_2O_5, the yield increased to approximately 6.1 to 7.6 Mg ha^{-1} without O_2 fertilizer and 7.1 to 8.4 Mg ha^{-1} with O_2 fertilizer. The greatest yields were observed at 135 kg ha^{-1} P_2O_5, where yields peaked at around 6.8 to 8.6 Mg ha^{-1} without O_2 fertilizer and 7.2 to 8.7 Mg ha^{-1} with O_2 fertilizer. Similar results were found in fall 2023 (Figure 4B), where the control treatment also exhibited the lowest yields, ranging from 3.2 to 4.4 Mg ha^{-1}. At 45 kg ha^{-1} P_2O_5, the yield increased to approximately 3.9 to 6.5 Mg ha^{-1} without O_2 fertilizer and 5.1 to 7.2 Mg ha^{-1} with O_2 fertilizer. The pod yield continued to rise with increasing P_2O_5 rates, reaching around 5.3 to 5.9 Mg ha^{-1} without O_2 fertilizer and 5.6 to 7.1 Mg ha^{-1} with O_2 fertilizer at 90 kg ha^{-1}. The greatest yields were observed at 135 kg ha^{-1} P_2O_5, where the yield peaked at approximately 6.8 to 7.4 Mg ha^{-1} without O_2 fertilizer and 6.8 to 8.9 Mg ha^{-1} with O_2 fertilizer. At the greatest rate of 179 kg ha^{-1} P_2O_5, the yield slightly decreased to an average 6.5 Mg ha^{-1} without O_2 fertilizer but remained relatively high at 7.2 Mg ha^{-1} with O_2 fertilizer. Comparatively, pod yields were generally higher in fall 2022 than in fall 2023 across all P_2O_5 rates, with the application of O_2 fertilizer consistently enhancing yields in both years. The results indicate that the combination of P_2O_5 and O_2 fertilizer significantly influences pod yield, with the maximum yields achieved at 135 kg ha^{-1} P_2O_5 in both growing seasons.

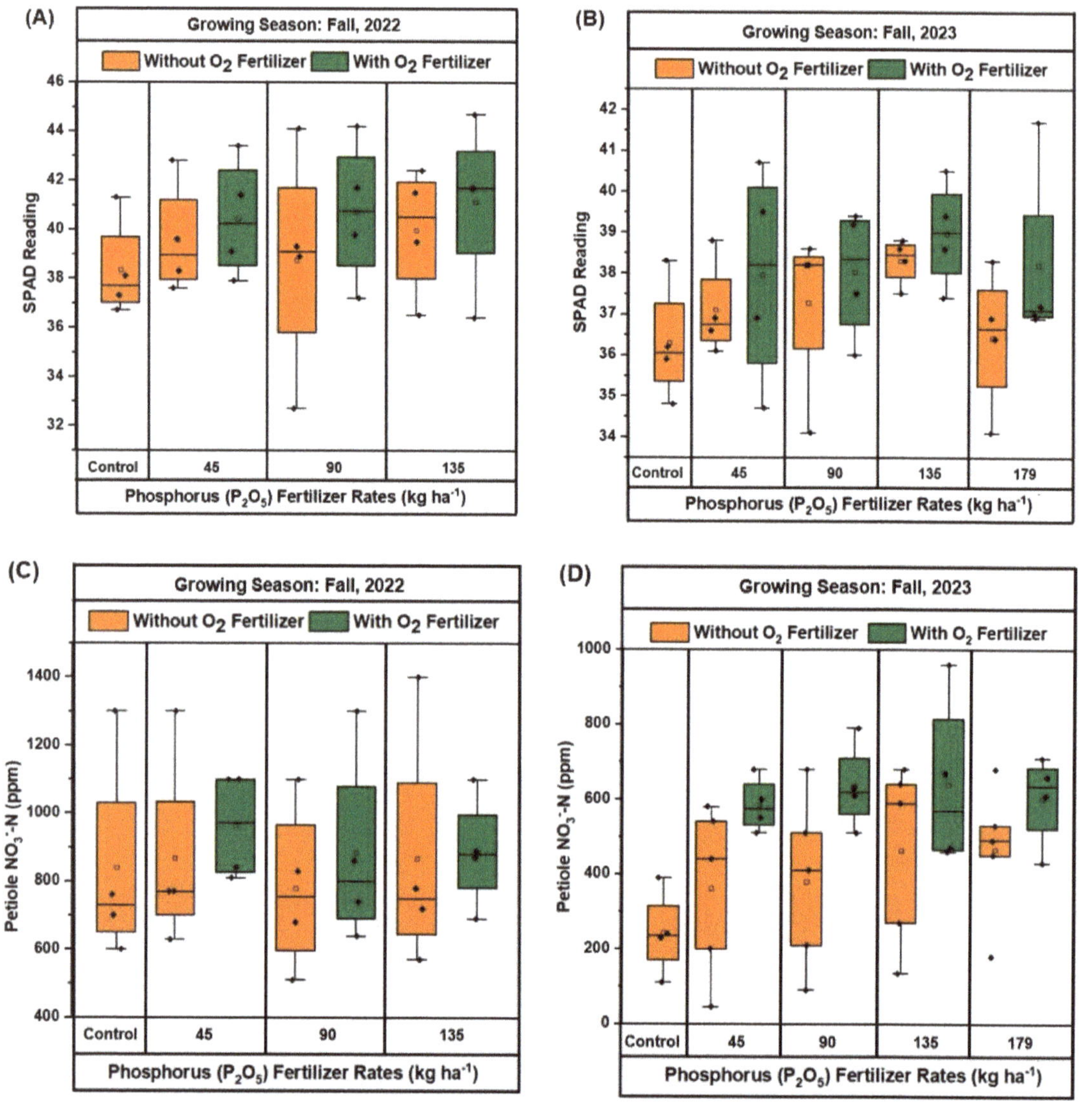

Figure 3. Effect of oxygen (O_2) fertilizer on the SPAD value (**A**,**B**) and petiole NO_3^--N concentration (**C**,**D**) on green beans grown with different rates of phosphorus (P_2O_5) fertilization during fall 2022 (**A**,**C**) and 2023 (**B**,**D**) at Hastings. In all boxplots, black dots represent the averages for each treatment. Boxplots show the median (the line that divides the box into two parts) and the upper and lower quantiles (the ends of the box). The whiskers of the boxplots indicate the variability outside the lower and upper quartiles (Q), being calculated as (Q1–1.5(Q3–Q1)) and (Q3–1.5(Q3–Q1)), respectively.

The boxplots in Figure 5A–F illustrate the effects of different P_2O_5 rates, with and without O_2 fertilizer, on yield attributes of green beans grown in the season fall 2022 and fall 2023. In fall 2022 (Figure 5A), the control treatment had the lowest pod length, ranging from 122 to 126 mm. At 45 kg ha^{-1} P_2O_5, the pod length ranged from 119 to 126 mm without O_2 fertilizer and 127 to 129 mm with O_2 fertilizer, with the greatest pod length observed at 135 kg ha^{-1} P_2O_5. In fall 2023 (Figure 5B), the control treatment exhibited even lower pod lengths, ranging from 87 to 105 mm without O_2 fertilizer. The pod length increased consistently with higher P_2O_5 rates, peaking at 135 kg ha^{-1} P_2O_5, and slightly decreasing at 179 kg ha^{-1} P_2O_5. For pod volume, in fall 2022 (Figure 5C), the control treatment had the lowest values, ranging from 5 to 5.75 cm^3. As P_2O_5 rates increased, the pod volume showed an increase, peaking at 135 kg ha^{-1} P_2O_5. In fall 2023 (Figure 5D), the control treatment showed even lower pod volumes, ranging from 2.17 to 4.0 cm^3. The pod volume increased consistently with higher P_2O_5 rates, peaking at 135 kg ha^{-1} P_2O_5, and slightly decreasing at 179 kg ha^{-1} P_2O_5. Regarding pod dry weight, in fall 2022 (Figure 5E), the control treatment had the lowest values, and with increasing P_2O_5 rates, the pod dry weight increased, peaking at 135 kg ha^{-1} P_2O_5. In fall 2023 (Figure 5F), the control treatment showed even lower pod dry weights, with a consistent increase in pod dry weight with higher P_2O_5 rates, peaking at 135 kg ha^{-1} P_2O_5, and slightly decreasing at 179 kg ha^{-1} P_2O_5. These results indicate that both P_2O_5 and O_2 fertilizer positively influenced pod yield attributes, with the maximum values generally observed at 135 kg ha^{-1} P_2O_5 in both growing seasons.

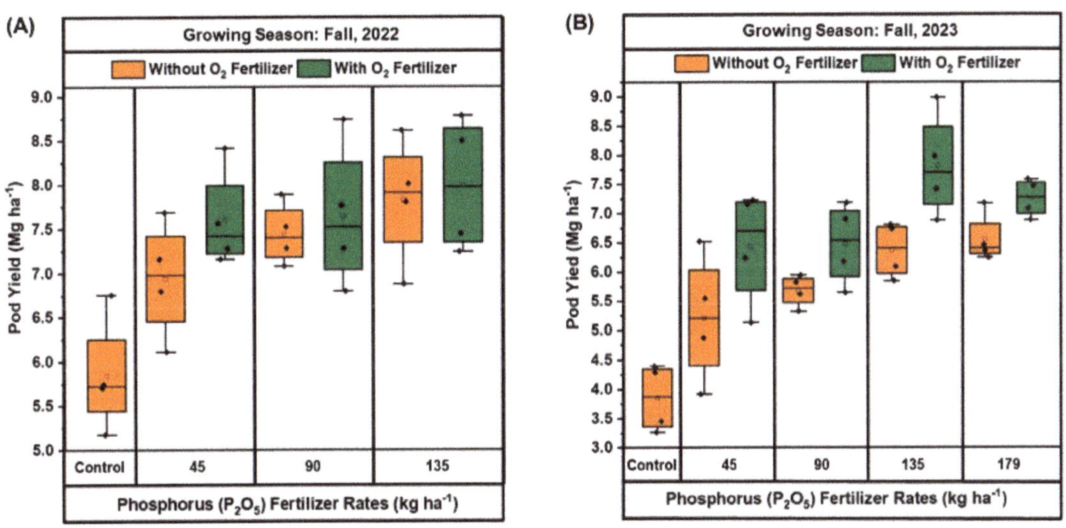

Figure 4. Comparison of snap bean yield under different rates of phosphorus (P_2O_5) fertilization, with and without oxygen (O_2) fertilizer, during fall 2022 (**A**) and 2023 (**B**) at Hastings. In all boxplots, black dots represent the average for each treatment. Boxplots show the median (the line that divides the box into two parts) and the upper and lower quantiles (the ends of the box). The whiskers of the boxplots indicate the variability outside the lower and upper quartiles (Q), being calculated as (Q1–1.5(Q3–Q1)) and (Q3–1.5(Q3–Q1)), respectively.

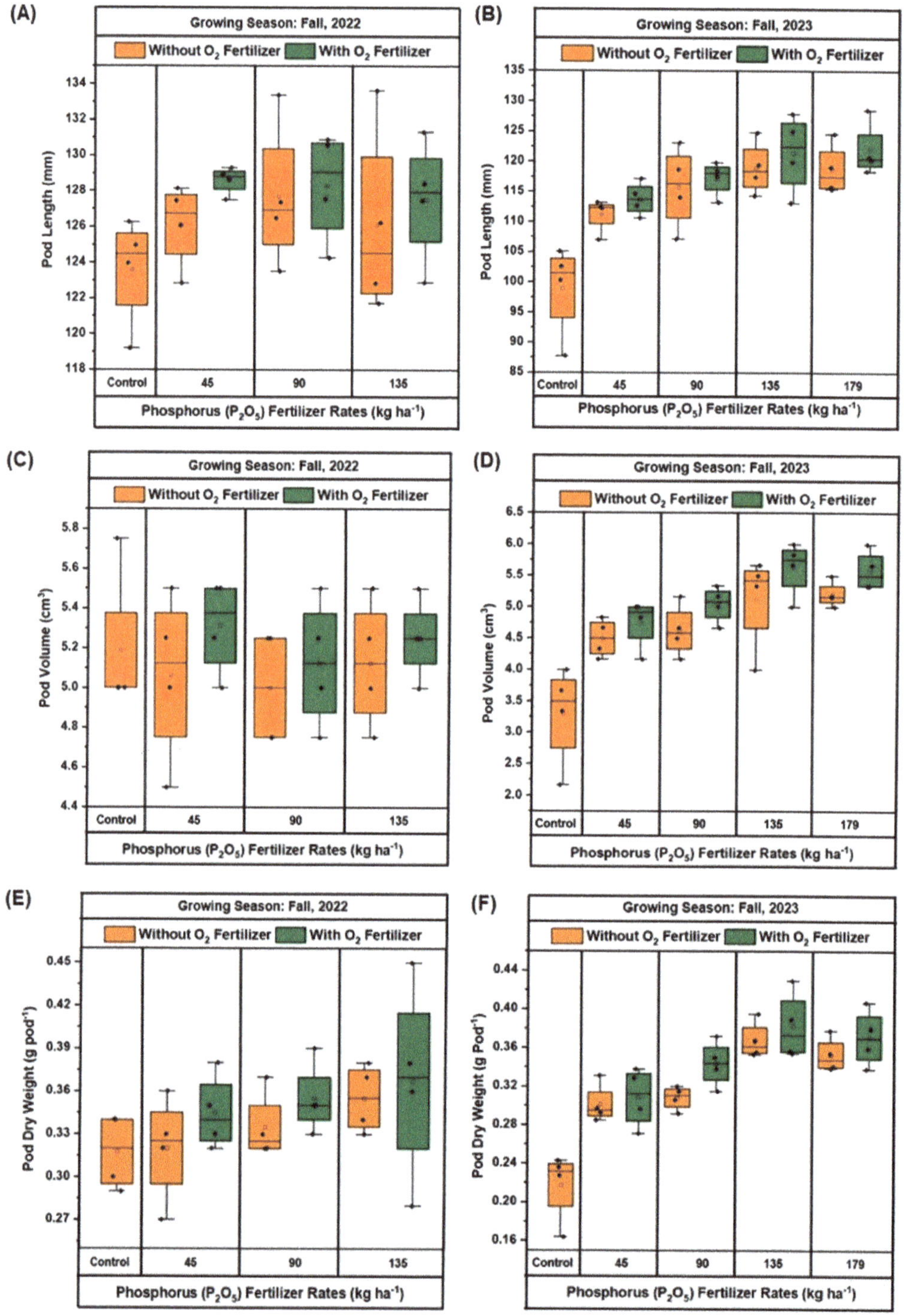

Figure 5. Effect of oxygen (O_2) fertilizer on pod length (**A**,**B**), pod volume (**B**,**C**), and pod dry weight (**E**,**F**) of green beans grown with different rates of phosphorus (P_2O_5) fertilization during fall 2022

(**A**,**C**,**E**) and 2023 (**B**,**D**,**F**) at Hastings. In all boxplots, black dots represent the averages for each treatment. Boxplots show the median (the line that divides the box into two parts) and the upper and lower quantiles (the ends of the box). The whiskers of the boxplots indicate the variability outside the lower and upper quartiles (Q), being calculated as (Q1–1.5(Q3–Q1)) and (Q3–1.5(Q3–Q1)), respectively.

2.4. Soil Available Phosphorus Concentrations

The dynamics of soil available P concentration were evaluated at 14, 28, and 42 days after sowing (DAS) across two growing seasons (fall 2022 and fall 2023) under different P fertilization rates and O_2 fertilization treatments (Figure 6A,B). In fall 2022 (Figure 6A), the control group soil without O_2 fertilizer showed an average median available P concentration of 48.34 mg kg^{-1}. However, with O_2 fertilizer, the available P concentration slightly increased. At the 45 kg ha^{-1} P rate, available P concentrations without O_2 fertilizer were lower, with a slight increase observed at 28 DAS and a decrease by 42 DAS. In contrast, with O_2 fertilizer, the available P concentration was higher at 14 DAS and decreased over time. For the 90 kg ha^{-1} P treatment, available P concentrations showed variability, with a maximum at 14 DAS and a gradual decrease by 42 DAS, more pronounced with O_2 fertilizer. At the 135 kg ha^{-1} P_2O_5 rate, the maximum available P concentration was 50.01 mg kg^{-1} at 14 DAS with O_2 fertilizer and decreased over time, whereas without O_2 fertilizer, the concentrations were higher and decreased over the period.

In fall 2023 (Figure 6B), the available P concentrations were notably higher across all treatments compared to fall 2022. For the control group soil, available P concentrations ranged from 64.26 mg kg^{-1} without O_2 fertilizer, with a gradual decrease over time. At the 45 kg ha^{-1} phosphorus rate, the available P concentration was maximum at 14 DAS, and then decreased by 42 DAS, with higher concentrations observed with O_2 fertilizer. For the 90 kg ha^{-1} phosphorus treatment, a similar trend was observed, with the maximum available P concentration being at 14 DAS and gradually decreasing, more pronounced with O_2 fertilizer. At the greatest rates of 135 kg ha^{-1} and 179 kg ha^{-1} phosphorus, the available P concentrations were greatest at 14 and decreased significantly by 42 DAS, showing the highest variability and greatest initial concentrations with O_2 fertilization. Overall, the results demonstrate that soil available P concentrations were significantly higher with the application of O_2 fertilizer and tended to decrease over time. This trend was consistent across both growing seasons and all P_2O_5 application rates, indicating the dynamic nature of P availability influenced by P and O_2 fertilization.

2.5. Crop Phosphorus Accumulation

The P uptake in the above-ground parts of green beans varied with different P_2O_5 rates and O_2 fertilization treatments across two growing seasons (Figure 7A,B). In fall 2022 (Figure 7A), the control group without O_2 fertilizer exhibited an average median P uptake of 35.24 mg plant^{-1}. In contrast, the presence of O_2 fertilizer substantially increased P uptake across all phosphorus rates. For instance, at the 45 kg ha^{-1} phosphorus rate, P uptake rose to about 60 mg plant^{-1} with O_2 fertilizer compared to 45 mg plant^{-1} without it. The 90 kg ha^{-1} phosphorus treatment showed an increase in P uptake from around 50 mg plant^{-1} without O_2 fertilizer to approximately 72.05 mg plant^{-1} with it. At the greatest P_2O_5 rate of 135 kg ha^{-1}, P uptake reached about 63.88 mg plant^{-1} without O_2 fertilizer and further increased to 76.31 mg plant^{-1} with O_2 fertilizer.

In fall 2023 (Figure 7B), the overall P uptake was lower compared to fall 2022 across all treatments. The control group without O_2 fertilizer had an average median P uptake of 15.5 mg plant^{-1}. At the 45 kg ha^{-1} P_2O_5 rate, P uptake increased from an average of 22.9 mg plant^{-1} without O_2 fertilizer to about 25.9 mg plant^{-1} with it. A similar trend was observed at the 90 kg ha^{-1} P_2O_5 treatment, where P uptake rose from around 28.69 mg plant^{-1} without O_2 fertilizer to 30.8 mg plant^{-1} with it. For the 135 kg ha^{-1} P_2O_5 rate, P uptake reached around 30.2 mg plant^{-1} without O_2 fertilizer and was

32.9 mg plant^{-1} with it. At the greatest P_2O_5 rate of 179 kg ha^{-1}, P uptake was approximately 33.99 mg plant^{-1} without O_2 fertilizer and slightly increased with O_2 fertilizer. These results indicate that O_2 fertilization consistently enhances P uptake in the aboveground parts of green beans, with more pronounced effects observed in fall 2022 compared to fall 2023.

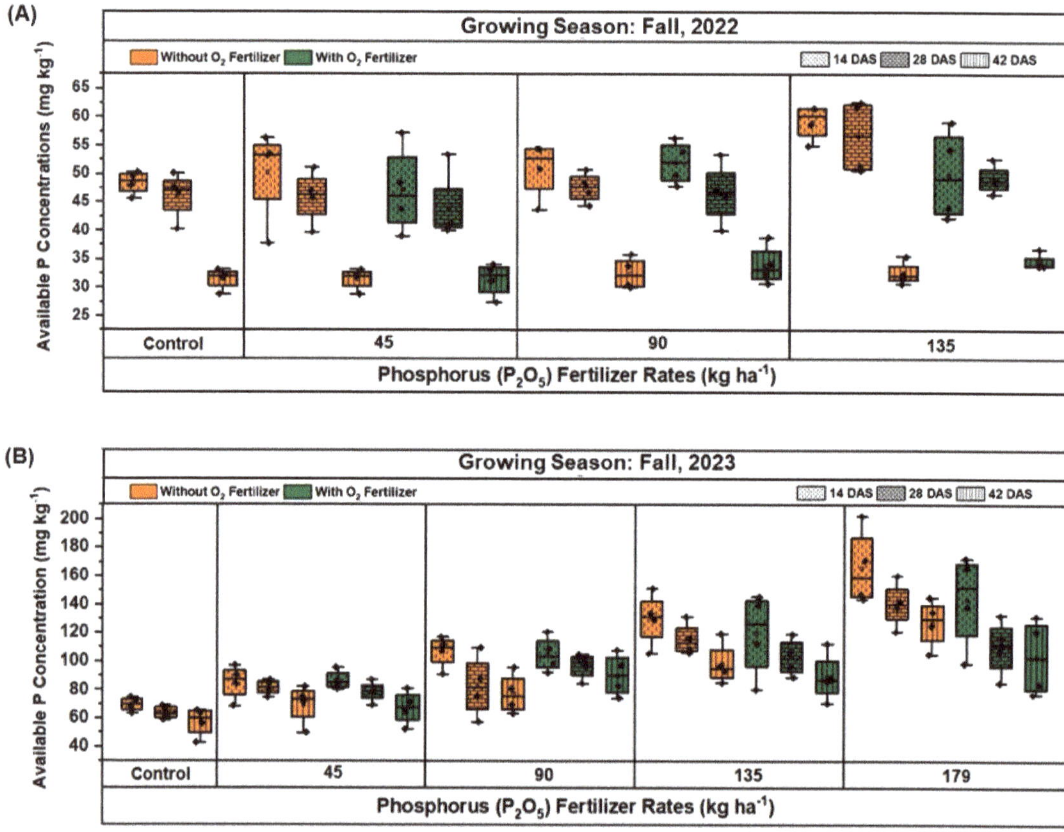

Figure 6. Comparison of soil available phosphorus (P) concentration under different rates of phosphorus (P_2O_5) fertilization, with and without oxygen (O_2) fertilizer in different growth stages during fall 2022 (**A**) and 2023 (**B**) at Hastings. In all boxplots, black dots represent the averages for each treatment. Boxplots show the median (the line that divides the box into t parts) and the upper and lower quantiles (the ends of the box). The whiskers of the boxplots indicate the variability outside the lower and upper quartiles (Q), being calculated as (Q1–1.5(Q3–Q1)) and (Q3–1.5(Q3–Q1)), respectively.

2.6. Correlation and Principal Component Analysis

The correlation analysis highlights strong positive correlations (indicated in red) among key crop variables such as dry biomass, SPAD, petiole NO_3^--N, pod yield, pod length, pod volume, pod dry weight, and plant nutrient contents (N, P, K, Ca, Mg), with significant correlations at $p < 0.01$, indicating these parameters tend to increase together (Figure 8). Specifically, a higher dry biomass is strongly associated with increased SPAD, petiole NO_3^--N, pod yield, pod length, pod volume, and pod dry weight. Plant nutrient contents also show strong positive correlations with each other and with pod parameters. On the other hand, negative correlations (indicated in blue) are observed with varying degrees between plant variables and certain soil properties such as soil P, K, Ca, Mg, pH, and CEC. The negative correlation of soil pH and CEC with several plant parameters suggests that higher soil pH and CEC levels may negatively impact plant growth. Overall, the analysis underscores the importance of balanced soil nutrient management for optimal crop performance.

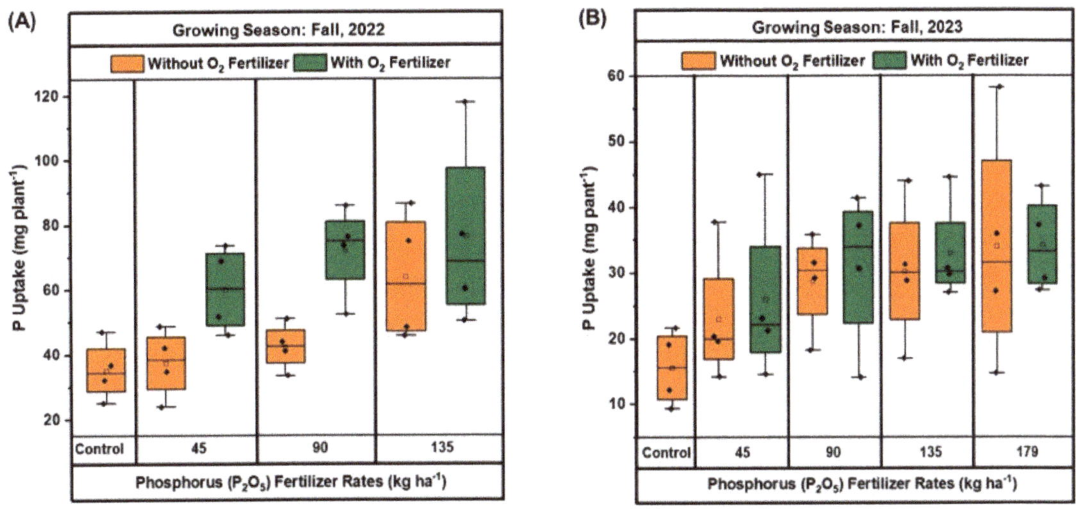

Figure 7. Effect of oxygen (O_2) fertilizer on phosphorus (P) uptake in green beans grown with different rates of phosphorus (P_2O_5) fertilization during fall 2022 (**A**) and 2023 (**B**) at Hastings. In all boxplots, black dots represent the averages for each treatment. Boxplots show the median (the line that divides the box into two parts) and the upper and lower quantiles (the ends of the box). The whiskers of the boxplots indicate the variability outside the lower and upper quartiles (Q), being calculated as (Q1–1.5(Q3–Q1)) and (Q3–1.5(Q3–Q1)), respectively.

The PCA score plot (Figure 9A) and loading plot (Figure 9B) provide insights into the variation and relationships among the crop and soil variables. In the score plot (Figure 9A), data points from fall 2022 and fall 2023 are clearly separated along the first principal component (PC1), which explains 36.9% of the total variance, indicating distinct differences in crop and soil characteristics between the two years. The second principal component (PC2) explains an additional 16.6% of the variance. The PCA loading plot (Figure 9B) reveals the contribution of multiple variables to these principal components. Plant growth parameters such as dry biomass, SPAD, petiole NO_3^--N, pod yield, pod length, pod volume, pod dry weight, and plant nutrients (N, P, K, Ca, and Mg) are positively associated with PC1, suggesting these factors are major contributors to the variance in the data. Conversely, soil properties such as soil P, Ca, Mg, pH, and CEC show significant positive loadings on PC2, while soil K negatively influences PC2. These results indicate that plant growth parameters and nutrient contents are the primary factors differentiating the samples along PC1, whereas soil properties significantly influence the variance captured by PC2. This analysis underscores the importance of both plant and soil characteristics in understanding the variability in the data across different growing seasons.

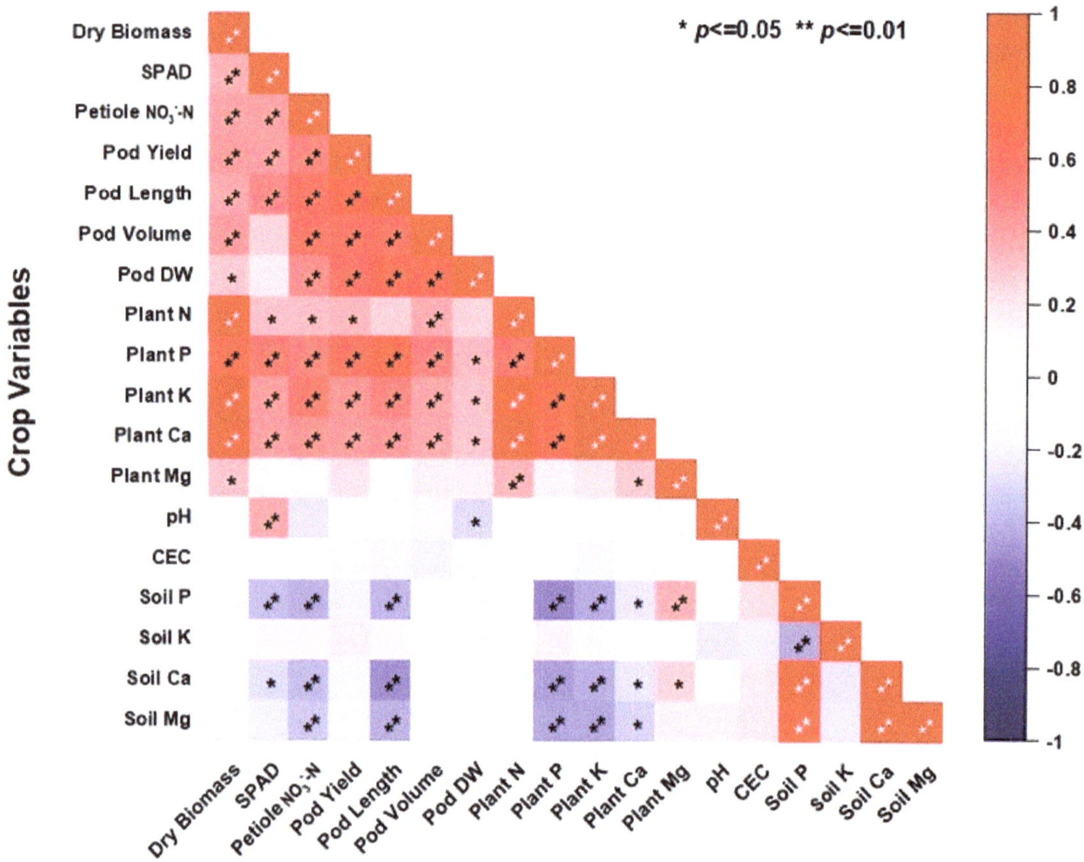

Figure 8. Pearson's correlation matrix showing the correlation (r) among different plant and soil properties along with p-values for the snap beans treated with different doses of phosphorus and oxygen fertilizer during fall 2022 and 2023 at Hastings. * = $p < 0.05$; ** = $p < 0.01$; SPAD = SPAD reading; pod DW = pod dry weight, plant N, P, K, Ca, and Mg = plant accumulated nitrogen, phosphorus, potassium, calcium, and magnesium concentration, respectively; soil P, K, Ca, and Mg = Mehlich-3 extractable soil phosphorus, potassium, calcium, and magnesium, respectively; pH = soil pH, CEC = soil cation exchange capacity.

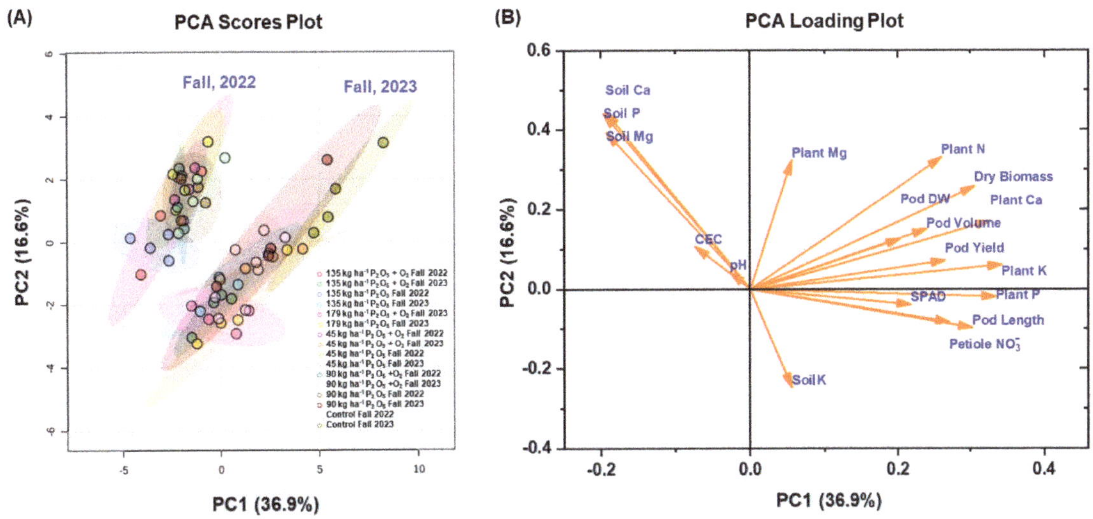

Figure 9. Principal component analysis (PCA) score plot (**A**) showing treatments (color points) and loading plot (**B**), indicating responses (orange arrows) for plant parameters and soil properties of snap beans treated with different doses of phosphorus and oxygen fertilizer during fall 2022 and 2023 at Hastings. FW = fresh weight; DW = dry weight; WC = water content; LN = leaf number; PL = plant height; SPAD = SPAD reading; tissue N, P, and K = tissue nitrogen, phosphorus, and potassium concentration, respectively; plant N, P, and K = plant accumulated nitrogen, phosphorus, and potassium, respectively; OM = organic matter; pH = soil pH; CEC = soil cation exchange capacity; soil P and K = Mehlich-3 extractable soil phosphorus and potassium, respectively.

3. Discussion

The findings from this two-year field study provide valuable insights into the impact of P and O_2 fertilization on snap bean growth, pod yield, and nutrient uptake. The study shows a clear positive correlation between the application of P and O_2 fertilizers and the improvement of biomass production (Figure 2) and plant growth parameters (Figure 3). In both fall 2022 and 2023, the combined application of 135 kg/ha of P and O_2 fertilizer resulted in the greatest biomass production compared to other treatments and overall plant growth parameters. Phosphorus application enhances petiole NO_3^--N content by improving N uptake and assimilation [36] and plant chlorophyll content or leaf greenness [37] by promoting efficient energy transfer during photosynthesis, thereby supporting overall plant growth and development. Our findings align with previous research that highlights the critical role of P in enhancing root development, biomass production, and overall plant growth [22]. Phosphorus significantly enhances plant chlorophyll content by promoting efficient energy transfer during photosynthesis, leading to improved overall plant health and growth. Additionally, the positive effect of O_2 fertilization on biomass production, as observed in this study, supports the hypothesis that supplemental O_2 can alleviate hypoxic stress in the root zone, thereby enhancing nutrient uptake and promoting healthier plant growth [38,39].

The results indicate that pod yield (Figure 4) and its related attributes (Figure 5), such as pod length, volume, and dry weight, are significantly influenced by the rate of P_2O_5 application and the presence of O_2 fertilizer. The greatest pod yield was consistently observed at 135 kg/ha of P_2O_5 combined with O_2 fertilizer in both growing seasons. This suggests a synergistic effect between P and O_2 fertilization, where enhanced root respiration due to O_2 fertilizer improved P uptake, resulting in better plant growth and greater yields. The reduction in pod yield in some cases at the greatest P_2O_5 rate (179 kg/ha) could be attributed to potential nutrient imbalances, which have been reported in previous studies [40]. When comparing yearly pod yield, we found that pod yields were generally higher in fall

2022 than in fall 2023 across all P_2O_5 application rates, possibly due to the significantly higher rainfall (Figure 1) during the pod maturation period in fall 2023. The consistent performance of the 135 kg/ha P_2O_5 rate across both years underscores the importance of optimizing fertilization rates to avoid such negative effects and maximize yield.

The study also highlights the dynamic nature of soil P availability and its interaction with O_2 fertilization. The application of O_2 fertilizer was found to increase soil available P concentrations, particularly at higher P_2O_5 rates. This effect was more pronounced in the first few weeks after sowing, suggesting that O_2 fertilization may enhance the early availability of P, which is crucial for root establishment and early plant growth [24]. The dynamics of soil P concentration also varied with the presence of O_2 fertilizer, with greater available P levels observed in treatments that received O_2 fertilizer. This finding indicates that O_2 fertilization may help maintain or even increase soil P availability over time, potentially reducing the need for excessive P inputs and mitigating environmental risks associated with P leaching and runoff [19,41].

The correlation (Figure 8) and PCA (Figure 9) conducted in this study provide further evidence of the strong association between plant growth parameters, pod yield, and nutrient content. The positive correlations among biomass, chlorophyll content or leaf greenness (SPAD), petiole NO_3^--N concentrations, and pod yield suggest that these factors are closely linked and collectively contribute to the overall productivity of the snap bean crop. Our results are supported by the findings of Rizzo et al. [42], who reported positive relationship between SPAD values, yield, and pod diameter. The PCA results indicate distinct differences between the two growing seasons, with plant growth parameters emerging as the primary drivers of variability. This underscores the impact of both year-specific environmental conditions and the combined effects of P and O_2 fertilization in determining crop performance. Each ellipse represents different treatment groups, highlighting that samples under similar treatments cluster together within each season, with tighter clusters indicating lower variability within groups. A key limitation of this study is the use of a single genotype, chosen to focus on the primary objective of evaluating the impact of P and O_2 fertilizers; however, future research will include multiple varieties to better understand genotype-specific responses to these treatments.

There is an urgent need to develop cost-effective and sustainable P fertilizer strategies that create a true win–win scenario for farmers. Such strategies should aim to improve agricultural production, reduce environmental pollution caused by P leaching and eutrophication, and lower costs for farmers. It is important to note that the O_2 fertilizer used in the current study has already been successfully commercialized by several agricultural innovation companies. This specific O_2 fertilizer is well-suited for agricultural applications due to its relative affordability. The cost of using O_2 fertilizer for snap beans, with a soil drench application method, is approximately \$64 per acre or \$160 per hectare. Based on the findings of this study, the use of O_2 fertilizer resulted in an average yield increase of approximately 1.1 Mg ha^{-1}, equivalent to a 20.4% increase over the 2022 and 2023 growing seasons. In 2023, the United States Department of Agriculture (USDA) reported that the average farm gate price of snap beans was \$60.80 per CWT [6], meaning the increased net profit from the use of O_2 fertilizer in this experiment could reach up to \$660 per hectare. Furthermore, field studies offer valuable opportunities for stakeholder engagement and collaboration. By involving farmers, agricultural extension services, regulatory authorities, and other industry stakeholders, field trials can facilitate knowledge exchange, address concerns, and promote the acceptance and adoption of O_2 fertilizer in agriculture.

4. Materials and Methods

4.1. Experimental Site Description

The experimental site was located at the University of Florida, Institute of Food and Agricultural Sciences (UF/IFAS), Hastings Agricultural Extension Center—specifically, the Cowpen Branch Demonstration and Research Facility situated in Hastings (29.690531 N, 81.441505 W), St. Johns County, FL, USA. The experiments were conducted at this station over two growing seasons, in fall 2022 and 2023, at the same field location. The experimental site experiences a subtropical climate throughout the year. During the fall, maximum temperatures generally range from approximately 23 to 31 °C, while minimum temperatures typically vary between 13 and 22 °C, accompanied by moderate humidity levels. Rainfall is evenly distributed across the seasons, with frequent thunderstorms and heavy rainfall occurring from late spring to early fall with an average annual precipitation of 1359.15 mm. The soil type at the experimental site is Spodosol (Ankona series), classified as acidic sandy soil with the following composition: sand (91.3%), silt (3%), and clay (5.7%) [43]. The chemical properties and critical limit of the 0–20 cm soil layer before the experiment were analyzed according to our previous report [44] and are shown in Table 1.

4.2. Experimental Design, Oxygen and Phosphorus Fertilizer Treatment

The experiments were arranged in a randomized complete block design with four blocks each with ten plots, four rows in each block, and a 12.2 m length for each plot. There was a 1.5 m skip between plots, and the total area of each block was 198.1 m^2. The trials were completed with hilled rows. The hilled rows (0.35 m in height) were formed with 1 m distance between row centers. Five P fertilizer levels were prepared and supplied as granular P_2O_5 (0-46-0) at rates of 0 (control), 45, 90, 135, and 179 kg ha^{-1}; around 45 kg ha^{-1} increments. In fall 2023, we added an additional P fertilizer rate based on the results of the fall 2022 experiment, as we found that 135 kg ha^{-1} still provided the maximum yield. Two rates of solid oxygen fertilizer, 0 and 45 kg ha^{-1}, were supplied using calcium peroxide (CaO_2). Each treatment was replicated four times. Before planting, different doses of P and O_2 fertilizers were applied to the soil surface of each row using a semi-automated, tractor-operated fertilizer applicator, and then pulverized and mixed into the soil.

4.3. Crop Production, Fertilizer, and Irrigation Management

The commercial green bean variety 'Caprice' (*Phaseolus vulgaris* L.) was used for this experiment in both years. 'Caprice' is a relatively high-yielding variety commonly cultivated by growers in Florida. Treated seeds of this variety were purchased from Clifton Seed Company (https://www.cliftonseed.com/ accessed on 6 May 2022). The seeds were sown in raised beds using a precision seed drill, with a row-to-row distance of 12.1 m and a plant-to-plant distance of 1.2 m, resulting in a plant density of 96,711 plants ha^{-1}. The seeds were planted on 13 September 2022 and 18 September 2023, respectively. The chronological timeline of soil and plant sampling, fertilizer application, and harvesting activities is provided in the Table 2. A weather station (UT30, Campbell Scientific Inc., Logan, UT, USA) from the Florida Automated Weather Network (FAWN—http://fawn.ifas.ufl.edu/) recorded site-specific weather conditions, including maximum and minimum soil and air temperatures, as well as precipitation (Figure 1). As advised by the UF/IFAS, plant protection and weed control procedures were followed. A seepage irrigation system was used in both years.

4.4. Crop-Growth-Related Data Collection

4.4.1. Crop Biomass Measurements

Plant samples for dry biomass analysis were collected from each plot at 7, 14, 21, 28, 35, 42, 49, and 56 days after planting (DAP) in both 2022 and 2023. During each sample day and harvest, shoot biomass was collected. The dry weights of the shoots were determined by weighing them after oven-drying at 65 °C until a consistent weight was attained.

4.4.2. Leaf Greenness Measurements

Leaf greenness was assessed prior to harvest using a Leaf Greenness Meter, Soil-Plant Analysis Development (SPAD-502 Plus, Konica Minolta, Japan). SPAD readings were taken from 30 newly fully developed leaves per plot, following the method outlined in our previous report [44].

Table 2. Chronological timeline of soil and plant sampling, fertilizer application, and harvesting activities during fall 2022 and fall 2023.

Date	Activity	Notes
06 July 2022	Soil sampling for physicochemical property analysis	4–8 composite samples
13 September 2022	P fertilizer treatment, N and K basal dose (40%), and planting	* UAN (32-0-0), TSP (0-46-0), blended PS + PC (0-0-55), N (40 kg ha^{-1}), and K_2O: N (40 kg ha^{-1})
20 September 2022	Plant tissue sampling	20 plants from each plot
27September 2022	Soil and plant tissue sampling	8–12 soil samples and 18 plants from each plot
04 October 2022	Plant tissue sampling	15 plants from each plot
11 October 2022	Soil and plant tissue sampling, remaining (60%) N and K application	8–12 soil samples and 12 plants from each plot. * UAN (32-0-0), blended PS + PC (0-0-55), N (60 kg ha^{-1}), and K_2O: N (60 kg ha^{-1})
18 October 2022	Plant tissue sampling	10 plants from each plot
25 October 2022	Soil and plant tissue sampling	8–12 soil sample and 8 plants from each plot
01 November 2022	Plant tissue sampling, SPAD and petiole NO_3^--N measurement	8 plants from each plot, 25 individual leaf samples for SPAD, and 10–12 NO_3^--N measurements
04 November 2022	Harvest	6 m from middle two rows
04 September 2023	Soil sampling for physicochemical property analysis	4–8 composite samples
18 September 2023	P fertilizer treatment, N and K basal dose (40%), and planting	* UAN (32-0-0), TSP (0-46-0), blended PS + PC (0-0-55), N (40 kg ha^{-1}), and K_2O: N (40 kg ha^{-1})
25 September 2023	Plant tissue sampling	20 plants from each plot
02 October 2023	Soil and plant tissue sampling	8–12 soil samples and 18 plants from each plot
09 October 2023	Plant tissue sampling	15 plants from each plot
16 October 2023	Soil and plant tissue sampling, remaining (60%) N and K application	8–12 soil samples and 12 plants from each plot. * UAN (32-0-0), blended PS + PC (0-0-55), N (60 kg ha^{-1}), and K_2O: N (60 kg ha^{-1})
23 October 2023	Plant tissue sampling	10 plants from each plot
30 October 2023	Soil and plant tissue sampling	8–12 soil samples and 8 plants from each plot
06 November 2023	Plant tissue sampling, SPAD and petiole NO_3^--N measurement	8 plants from each plot, 25 individual leaf samples for SPAD, and 10–12 NO_3^--N measurements
07 November 2023	Harvest	6 m from middle two rows

* UAN: urea ammonium nitrate, TSP: triple super phosphate, PS: potassium sulfate, and PC: potassium chloride.

4.4.3. Crop Nitrogen Uptake Measurements

Petiole sap nitrate-nitrogen (NO_3^--N) concentrations were measured prior to harvest with a Horiba LAQUAT Cardy nitrate meter. At least, 20 petioles were sampled for each measurement, depending on sap availability per petiole.

4.5. Pod Harvest and Pod-Yield-Related Quality Measurements

Snap bean plants were harvested on 4 November 2022, and 7 November 2023. Plants from the middle 6.1 m of the two central rows in each plot were harvested and transported to UF/IFAS, Horticultural Sciences Department, Gainesville. Pods were manually collected, weighed, and calculated for total yields. Pod length, volume, fresh and dry weight, and water content were measured from 30 randomly collected pods.

4.6. Nutrient Content Analysis

All plant and soil samples were sent to Waters Agricultural Laboratories (Camilla, GA, USA) for nutrient content analysis. Details of the plant and soil analysis procedure are given below:

4.6.1. Plant Nutrient Content Analysis

Following harvest, plant samples were dried at 65 °C and finely ground using a Thomas Wiley Laboratory mill (Model 4, Arthur H. Thomas Company, Philadelphia, PA, USA). Total N content was determined using the dry combustion method with a C/N Analyzer (LECO Elemental Analysis System, MO, USA) at 1200 °C. Homogenized samples (0.3 g) were placed in a digestion tube and digested with 5 mL concentrated nitric acid

(HNO_3). After cooling to room temperature, the digestion solution was transferred to a 25 mL volumetric flask, and the volume was adjusted using double-deionized water. Macronutrient (N, P, K, Ca, and Mg) concentrations were determined using inductively coupled plasma mass spectrometry (ICP-MS) (Agilent 7500a, Agilent Technologies, CA, USA), following a previously described method [45].

To ensure analytical accuracy, duplicate readings were recorded for every tenth sample. Additionally, a standard reference material from the National Institute of Standards and Technology (Gaithersburg, MD, USA) and a blank control (Digested HNO_3) were measured. Nutrient accumulation was calculated by multiplying the nutrient concentration by the dry tissue biomass.

4.6.2. Soil Nutrient Content Analysis

Soil samples from each plot were collected for soil nutrient dynamics analysis at 14, 28, and 42 days after planting (DAP) in both 2022 and 2023. Before the experiment, composite soil samples were analyzed (Table 1), while after harvest, analysis was conducted on bulk soil from each plot. The physicochemical properties of the experimental soils were assessed using standard laboratory procedures outlined in "Soil Sampling and Method of Analysis" [46]. Soil pH and electrical conductivity (EC) were measured using a dual-channel pH/mV/Ion/conductivity/DO meter (XL 600, Fischer Scientific, Pittsburgh, PA, USA) with a soil-to-water ratio of 1:5. Mehlich-3 extractable soil elements (P, K, Ca, Mg, Na, S, B, Mn, Zn, Fe, and Cu) were determined by extracting samples with Mehlich-3 (M3) solution at a soil-to-solution ratio of 1:10 [47] and analyzed using ICP-MS (Agilent 7500a, Agilent Technologies, Santa Clara, CA, USA).

For quality control, duplicate readings were taken for every tenth sample in each batch, and a standard reference material from the National Institute of Standards and Technology (Gaithersburg, MD, USA) was measured. Additionally, a blank control (M3 solution only) was utilized to ensure analytical accuracy.

4.7. Statistical Analysis of Data

Statistical analysis was performed with R [48] software (Version: 4.3.1), which was run through R Studio [49]. For each year, collected data underwent a one-way analysis of variance (ANOVA) for randomized complete block design through the *aov()* function. For multivariate analysis, data normalization, log transformation, and scaling were applied to reduce variance among variables, as the dataset included different plant and soil parameters. Pearson correlation matrices and principal component analysis (PCA) were then used to identify the principal components explaining the majority of variability in the dataset and to determine correlation (r) coefficients among various plant and soil parameters. Pearson correlation matrix tables and PCA were conducted using the corrplot [50] and factoMineR [51] packages, respectively. Figures were generated using OriginPro (https://www.originlab.com/).

5. Conclusions

The application of O_2 fertilizer offers a cost-effective and sustainable approach to improving P use efficiency and enhancing crop productivity while mitigating the environmental risks associated with excessive P inputs. This two-year field study demonstrated that the combined application of P and O_2 fertilizers significantly enhances snap bean growth, pod yield, and nutrient uptake. Specifically, the application of 135 kg ha^{-1} of phosphorus pentoxide (P_2O_5) alongside 45 kg ha^{-1} of calcium peroxide (CaO_2) yielded the greatest pod yield and biomass production, confirming a synergistic effect between P and O_2 that enhances root respiration, nutrient uptake, and overall crop performance. These findings indicate that the combined use of P and O_2 fertilizers is a promising strategy for maximizing yields and promoting sustainable nutrient management in snap bean cultivation, while also reducing the risks of phosphorus runoff and leaching. Correlation and principal component analyses further underscore strong associations between plant growth parameters, nutrient content, and yield, highlighting year-specific environmental factors as key contributors to variability in crop performance. Overall, these insights provide practical guidance for farmers and contribute to filling the knowledge gap regarding the combined use of P and O_2 fertilization in field conditions.

Author Contributions: Conceptualization, G.L., M.C.G. and M.J.I.S.; methodology, G.L., M.C.G. and M.J.I.S.; formal analysis, M.J.I.S.; investigation, M.C.G., M.J.I.S. and E.M.S.; resources, G.L., M.C.G. and M.J.I.S.; validation, M.J.I.S. and G.L.; data curation, M.C.G. and M.J.I.S.; writing—original draft preparation, M.J.I.S.; writing—review and editing, M.J.I.S., M.C.G., E.M.S. and G.L.; project administration, funding acquisition, and supervision, G.L.; All authors have read and agreed to the published version of the manuscript.

Funding: This work was financially supported by a grant (FDACS Contract 1/2022: 28897 and Contract 2/2023: 30171. Project 1/2022: P0279884 and Project 2/2023: P0310044) from Florida Department of Agriculture and Consumer Services.

Institutional Review Board Statement: Not applicable.

Data Availability Statement: The raw data generated for this study are available upon on request to the corresponding author.

Acknowledgments: We would like to express our cordial thanks to the personnel and leadership of UF/IFAS, Hastings Agricultural Extension Center for their support during this two years field study.

Conflicts of Interest: The authors declare that the research was conducted without any commercial or financial relationships that could be construed as a potential conflict of interest.

References

1. Menga, V.; Fares, C.; Campa, A.; Ferreira, J.J.; Bitocchi, E.; Papa, R.; Beleggia, R. Variability of Nutritional, Antioxidant, and Textural Traits of a Collection of Snap Beans of Different Colors. *Horticulturae* **2023**, *9*, 311. [CrossRef]
2. Dos Santos Neto, J.; Moda-Cirino, V.; Delfini, J.; Zeffa, D.M.; Nogueira, A.F.; Gonçalves, L.S.A. Morpho-agronomic and molecular characterization of snap bean diversity panel. *Genet. Resour. Crop Evol.* **2022**, *70*, 399–411. [CrossRef]
3. Lourenco, S.C.; Moldao-Martins, M.; Alves, V.D. Antioxidants of Natural Plant Origins: From Sources to Food Industry Applications. *Molecule* **2019**, *24*, 4132. [CrossRef]
4. Hosseinpour-Niazi, S.; Mirmiran, P.; Hedayati, M.; Azizi, F. Substitution of red meat with legumes in the therapeutic lifestyle change diet based on dietary advice improves cardiometabolic risk factors in overweight type 2 diabetes patients: A cross-over randomized clinical trial. *Eur. J. Clin. Nutr.* **2015**, *69*, 592–597. [CrossRef]
5. Food and Agricultural Organization (FAO). *World Food and Agriculture—Statistical Yearbook 2023*; Food and Agricultural Organization (FAO): Rome, Italy, 2023.
6. United States Department of Agriculture (USDA). *Crop Production 2023 Summary, January*; United States Department of Agriculture (USDA): Washington, DC, USA, 2024.
7. Yuan, M.; Ruark, M.D.; Bland, W.L.J.F.C.R. A simple model for snap bean (*Phaseolus vulgaris* L.) development, growth and yield in response to nitrogen. *Field Crops Res.* **2017**, *211*, 125–136. [CrossRef]
8. Vance, C.P.; Uhde-Stone, C.; Allan, D.L. Phosphorus acquisition and use: Critical adaptations by plants for securing a nonrenewable resource. *New Phytol.* **2003**, *157*, 423–447. [CrossRef]
9. Hernandez, I.; Munne-Bosch, S. Linking phosphorus availability with photo-oxidative stress in plants. *J. Exp. Bot.* **2015**, *66*, 2889–2900. [CrossRef] [PubMed]
10. Reich, P.S. AW Role of phosphorus and nitrogen in photosynthetic and whole plant carbon gain and nutrient use efficiency in eastern white pine. *J. Oecologia* **1988**, *77*, 25–33. [CrossRef]
11. Marschner, H. *Marschner's Mineral Nutrition of Higher Plants*; Academic Press: Cambridge, MA, USA, 2011.
12. Dhillon, J.; Torres, G.; Driver, E.; Figueiredo, B.; Raun, W.R. World Phosphorus Use Efficiency in Cereal Crops. *Agron. J.* **2017**, *109*, 1670–1677. [CrossRef]
13. Zou, T.; Zhang, X.; Davidson, E.A. Global trends of cropland phosphorus use and sustainability challenges. *Nature* **2022**, *611*, 81–87. [CrossRef]
14. Jagdeep, S.; Gobinder, S.; Gupta, N. Balancing phosphorus fertilization for sustainable maize yield and soil test phosphorus management: A long-term study using machine learning. *Field Crops Res.* **2023**, *304*, 109169. [CrossRef]
15. Liao, X.; Liu, G.; Hogue, B.; Li, Y.; Nicholson, F. Phosphorus availability and environmental risks in potato fields in North Florida. *Soil Use Man.* **2015**, *31*, 308–312. [CrossRef]
16. Ran, X.; Deng, Y.; Uppuluri, N.S.T.; Li, B.; Zheng, Y.; Chen, P.; Dong, R.; Muller, J.; Guo, J.; Oechsner, H. Hotspots and future trends of phosphorus recycling from livestock manure: A bibliometric review. *Sci. Total Environ.* **2023**, *892*, 164346. [CrossRef]

17. Yang, X.; Liu, C.; Liang, C.; Wang, T.; Tian, J. The Phosphorus-Iron Nexus: Decoding the Nutrients Interaction in Soil and Plant. *Int. J. Mol. Sci.* **2024**, *25*, 6992. [CrossRef]
18. Zhang, W.; Zhang, W.; Wang, X.; Liu, D.; Zou, C.; Chen, X. Quantitative evaluation of the grain zinc in cereal crops caused by phosphorus fertilization. A meta-analysis. *Agron. Sustain. Dev.* **2021**, *41*, 6. [CrossRef]
19. Carpenter, S.R. Phosphorus control is critical to mitigating eutrophication. *Proc. Natl. Acad. Sci. USA* **2008**, *105*, 11039–11040. [CrossRef]
20. Sharpley, A.; Jarvie, H.P.; Buda, A.; May, L.; Spears, B.; Kleinman, P. Phosphorus legacy: Overcoming the effects of past management practices to mitigate future water quality impairment. *J. Environ. Qual.* **2013**, *42*, 1308–1326. [CrossRef]
21. Graham, P.H.; Vance, C.P. Legumes: Importance and constraints to greater use. *Plant Physiol.* **2003**, *131*, 872–877. [CrossRef]
22. Youssef, S.M.; Riad, G.S.; Abd Elhady, S.A. Effect of Phosphorus Sources and Arbuscular Mycorrhizal Inoculation on Growth and Productivity of Snap Bean (*Phaseolus vulgaris* L.). *Gesunde Pflanz.* **2017**, *69*, 139–148. [CrossRef]
23. Hinsinger, P. Bioavailability of soil inorganic P in the rhizosphere as affected by root-induced chemical changes: A review. *Plant Soil* **2001**, *237*, 173–195. [CrossRef]
24. Colmer, T.D. Long-distance transport of gases in plants: A perspective on internal aeration and radial oxygen loss from roots. *Plant Cell Environ.* **2003**, *26*, 17–36. [CrossRef]
25. Voesenek, L.A.; Sasidharan, R. Ethylene—And oxygen signalling—Drive plant survival during flooding. *Plant Biol.* **2013**, *15*, 426–443. [CrossRef] [PubMed]
26. Armstrong, W.; Drew, M.C. Root growth and metabolism under oxygen deficiency. In *Plant Roots*; CRC Press: Boca Raton, FL, USA, 2002; pp. 1139–1187.
27. Drew, M.C. Oxygen deficiency and root metabolism: Injury and Acclimation Under Hypoxia and Anoxia. *Annu. Rev. Plant Physiol. Plant Mol. Biol.* **1997**, *48*, 223–250. [CrossRef] [PubMed]
28. Mei, J.; Wang, W.; Peng, S.; Nie, L. Seed Pelleting with Calcium Peroxide Improves Crop Establishment of Direct-seeded Rice under Waterlogging Conditions. *Sci. Rep.* **2017**, *7*, 4878. [CrossRef] [PubMed]
29. Wang, Z.; Han, Y.; Luo, S.; Rong, X.; Song, H.; Jiang, N.; Li, C.; Yang, L. Calcium peroxide alleviates the waterlogging stress of rapeseed by improving root growth status in a rice-rape rotation field. *Front. Plant Sci.* **2022**, *13*, 1048227. [CrossRef] [PubMed]
30. Pang, J.; Ryan, M.H.; Lambers, H.; Siddique, K.H. Phosphorus acquisition and utilisation in crop legumes under global change. *Curr. Opin. Plant Biol.* **2018**, *45*, 248–254. [CrossRef] [PubMed]
31. Mitran, T.; Meena, R.S.; Lal, R.; Layek, J.; Kumar, S.; Datta, R. Role of soil phosphorus on legume production. In *Legumes for Soil Health and Sustainable Management*; Springer: Berlin/Heidelberg, Germany, 2018; pp. 487–510.
32. Negash, A.; Solomon, T.; Getachew, E.J.I.J. Growth of snap bean (*Phaseolus vulgaris* L.) as affected by N and P fertilizer Rates at Jimma, Southwestern Ethiopia. *Inter. J. Res. Agric. Sci.* **2017**, *4*, 2348–3997.
33. Liu, D.; Paul, A.L.; Morgan, K.T.; Liu, G. Effects of oxygen fertilization on damage reduction in flooded snap bean (*Phaseolus vulgaris* L.). *Sci. Rep.* **2022**, *12*, 4282. [CrossRef]
34. Liu, G.; Porterfield, D.M. Oxygen enrichment with magnesium peroxide for minimizing hypoxic stress of flooded corn. *J. Plant Nutr. Soil Sci.* **2014**, *177*, 733–740. [CrossRef]
35. Dorau, K.; Wessel-Bothe, S.; Milbert, G.; Schrey, H.P.; Elhaus, D.; Mansfeldt, T.J.C. Climate change and redoximorphosis in a soil with stagnic properties. *Cate* **2020**, *190*, 104528. [CrossRef]
36. Zhao, J.; Huang, R.; Yang, K.; Ma, C.; Zhang, Q.J.A. Effects of nitrogen and phosphorus fertilization on photosynthetic properties of leaves and agronomic characters of Alfalfa over three consecutive years. *Agriculture* **2022**, *12*, 1187. [CrossRef]
37. Wu, C.; Wang, Z.; Sun, H.; Guo, S. Effects of different concentrations of nitrogen and phosphorus on chlorophyll biosynthesis, chlorophyll a fluorescence, and photosynthesis in *Larix olgensis* seedlings. *Front. For. China* **2006**, *1*, 170–175. [CrossRef]
38. Fu, X.; Xu, Y.J.S. Dynamic metabolic changes in arabidopsis seedlings under hypoxia stress and subsequent reoxygenation recovery. *Stresses* **2023**, *3*, 86–101. [CrossRef]
39. Planchet, E.; Lothier, J.; Limami, A.M. Hypoxic respiratory metabolism in plants: Reorchestration of nitrogen and carbon metabolisms. In *Plant Respiration: Metabolic Fluxes and Carbon Balance*; Springer: Berlin/Heidelberg, Germany, 2017; pp. 209–226.
40. Singh, J.; Brar, B.S.; Sekhon, B.S.; Mavi, M.S.; Singh, G.; Kaur, G.J.N.C.i.A. Impact of long-term phosphorous fertilization on Olsen-P and grain yields in maize–wheat cropping sequence. *Nutr. Cyc. Agroecosystem* **2016**, *106*, 157–168. [CrossRef]
41. Sharpley, A.N.; Kleinman, P.J.; Flaten, D.N.; Buda, A.R. Critical source area management of agricultural phosphorus: Experiences, challenges and opportunities. *Water Sci. Technol.* **2011**, *64*, 945–952. [CrossRef] [PubMed]
42. Rizzo, G.F.; Al Achkar, N.; Treccarichi, S.; Malgioglio, G.; Infurna, M.G.; Nigro, S.; Tribulato, A.; Branca, F.J.A. Use of bioinoculants affects variation in snap bean yield grown under deficit irrigation. *Agriculture* **2023**, *13*, 865. [CrossRef]
43. Mylavarapu, R.; Harris, W.; Hochmuth, G. Agricultural Soils of Florida. SL441 Gainesville: University of Florida Institute of Food and Agricultural Sciences 2022. Available online: https://edis.ifas.ufl.edu/publication/SS655 (accessed on 13 October 2024).
44. Shohag, M.J.I.; Tian, S.; Sriti, N.; Liu, G. Enhancing Bok Choy growth through synergistic effects of hydrogel and different nitrogen fertilizer forms. *Sci. Hortic.* **2024**, *336*, 113400. [CrossRef]
45. Wu, C.Y.; Feng, Y.; Shohag, M.J.I.; Lu, L.; Wei, Y.Y.; Gao, C.; Yang, X.E. Characterization of 68 Zn uptake, translocation, and accumulation into developing grains and young leaves of high Zn-density rice genotype. *J. Zhejiang Univ. Sci. B* **2011**, *12*, 408–418. [CrossRef] [PubMed]
46. Carter, M.R.; Gregorich, E.G. *Soil Sampling and Methods of Analysis*; CRC Press: Boca Raton, FL, USA, 2007.

47. Mehlich, A. Mehlich 3 soil test extractant: A modification of Mehlich 2 extractant. *Commun. Soil Sci. Plant Anal.* **1984**, *15*, 1409–1416. [CrossRef]
48. R Core Team. R: A Language and Environment for Statistical Computing (Version 4.3.1). R Foundation for Statistical Computing, Vienna, Austria. 2023. ULR. Available online: https://www.r-project.org/ (accessed on 12 August 2024).
49. RStudio Team. RStudio: Integrated Development Environment for R. RStudio, PBC, Boston, MA. 2022. ULR. Available online: https://posit.co/ (accessed on 13 October 2024).
50. Wei, T.; Simko, V. R Package "Corrplot": Visualization of a Correlation Matrix (Version 0.92). 2021 ULR. Available online: https://github.com/taiyun/corrplot (accessed on 13 October 2024).
51. Lê, S.; Josse, J.; Husson, F. FactoMineR: An R package for multivariate analysis. *J. Stat. Sof.* **2008**, *25*, 1–18.

plants

MDPI

Article

Evaluating a Soil Amendment for Cadmium Mitigation and Enhanced Nutritional Quality in Faba Bean Genotypes: Implications for Food Safety

Liping Cheng [1], Jiapan Lian [1,2], Xin Wang [1], Mehr Ahmed Mujtaba Munir [1], Xiwei Huang [1], Zhenli He [3], Chengjian Xu [4], Wenbin Tong [4,*] and Xiaoe Yang [1,*]

1 Ministry of Education (MOE) Key Laboratory of Environmental Remediation and Ecosystem Health, College of Environmental and Resources Sciences, Zhejiang University, Hangzhou 310058, China; lpcheng@zju.edu.cn (L.C.); ljiapan@zju.edu.cn (J.L.); 22214128@zju.edu.cn (X.W.); muju212@mail.ustc.edu.cn (M.A.M.M.); 22114147@zju.edu.cn (X.H.)

2 State Key Laboratory for Conservation and Utilization of Subtropical Agri-Biological Resources, Guangxi University, Nanning 530004, China

3 Department of Soil, Water and Ecosystem Sciences, Indian River Research and Education Center, University of Florida—IFAS, Fort Pierce, FL 34945, USA; zhe@ufl.edu

4 Qujiang District Agricultural Technology Extension Center, Quzhou 324022, China; m13819005727@163.com

* Correspondence: zjqztwb@163.com (W.T.); xeyang@zju.edu.cn (X.Y.); Tel./Fax: +86-0570-8753680 (W.T.); +86-0571-88982907 (X.Y.)

Abstract: Soil amendments combined with low cadmium (Cd)-accumulating crops are commonly used for remediating Cd contamination and ensuring food safety. However, the combined effects of soil amendments and the cultivation of faba beans (*Vicia faba* L.)—known for their high nutritional quality and low Cd accumulation—in moderately Cd-contaminated soils remain underexplored. This study investigates the impact of a soil amendment (SA) on agronomic traits, seed nutrition, and Cd accumulation in 11 faba bean genotypes grown in acidic soil (1.3 mg·kg^{-1} Cd, pH 5.39). The SA treatment increased soil pH to 6.0 (an 11.31% increase) and reduced DTPA-Cd by 37.1%. Although the average yield of faba beans decreased marginally by 8.74%, it remained within the 10% national permissible limit. Notably, SA treatment reduced Cd concentration in seeds by 60% and significantly mitigated Mn and Al toxicity. Additionally, SA treatment enhanced levels of essential macronutrients (Ca, Mg, P, S) and micronutrients (Mo, Cu) while lowering Phytate (Phy)/Ca, Phy/Mg, and Phy/P ratios, thus improving mineral nutrient bioavailability. Among the genotypes, F3, F5, and F6 showed the most favorable balance of nutrient quality, and yield following SA application. This study provides valuable insights into the effectiveness of SA for nutrient fortification and Cd contamination mitigation in Cd-contaminated farmland.

Keywords: cadmium contamination; soil remediation; *Vicia faba* L.; agronomic traits; food security

1. Introduction

Cadmium (Cd) is one of the most common and highly toxic heavy metal pollutants in agricultural ecosystems, with a half-life of 10–30 years. Cadmium accumulates in the human liver, kidneys, lungs, and bones and poses serious health risks. The International Agency for Research on Cancer has classified Cd as a Group 1 carcinogen [1]. Cadmium primarily enters the human body through the soil–crop–human chain, making soil contamination with Cd a major threat to food safety. In China, industrial emissions, agricultural activities, and climate change collectively contribute to soil Cd pollution [2–5]. The 2014 China National Soil Pollution Survey Report revealed that 16.1% of soils have excessive heavy metal content, with Cd exceeding the standard in 7.0% of cases [6,7]. Approximately 13 million hectares of farmland in China, particularly in industrialized regions such as Hunan, Guizhou, and Nanning, are affected by Cd pollution [6,7]. Yang et al. predicted through machine learning that by 2040, 41% of China's population will be exposed to Cd levels that exceed the safety thresholds set by European and American food safety agencies [8,9]. Therefore, strategies for effective management of Cd pollution are imperative.

Soil Cd pollution can be controlled via phytoremediation, application of soil passivation agents, agronomic measures, and bioremediation [10–16]. However, reducing Cd accumulation in crops is a key strategy for minimizing human Cd exposure. Research indicates that the Cd accumulation capability of a crop is influenced by its genotype and soil conditions. Therefore, selecting low-Cd-accumulating crop genotypes combined with optimized soil management practices is an effective approach to reducing Cd pollution [17]. In recent years, the application of soil amendments (SAs) has become an important method for reducing Cd uptake by crops. Soil amendments alter soil properties, stabilize Cd levels, and regulate the bioavailability of nutrients, thereby promoting crop growth and mineral enrichment [10,18].

The faba bean (*Vicia faba* L.) is a versatile and productive legume known for its protein-rich seeds. Globally, faba beans are second only to soybeans in terms of production and have the highest protein content (29%) among legumes. Faba beans are highly tolerant to cold temperatures and are considered a viable alternative to soybeans [19]. China is one of the largest producers of faba beans, with the crop being often used as food and fodder. Faba beans are tolerant to waterlogging and temporary flooding, making them highly suitable for cereal and legume rotations [20,21], and for management schemes for Cd-contaminated farmland [22]. Faba bean genotypes exhibit notable differences in Cd absorption capacities, with some exhibiting a strong Cd tolerance [23–27]. However, research on the synergistic effects of faba bean genotypes and soil amendments, particularly in high-Cd-contaminated soils, remains limited. Systematic studies on the responses of different faba bean genotypes to soil amendments would thus provide novel solutions for Cd pollution management and food security.

In this study, we aimed to investigate the synergistic effects of an SA application and low-Cd-accumulating faba bean genotypes on Cd uptake, translocation, and accumulation; agronomic traits; and seed nutritional quality. The findings of this study are expected to provide insights into sustainable agricultural practices that enhance food safety and crop nutrition in Cd-contaminated farmlands and identify superior faba bean genotypes for use in such environments.

2. Results and Discussions

2.1. Agronomic Traits of Faba Bean Genotypes by Soil Amendment

2.1.1. Theoretical Yield of Faba Bean Genotypes

In field trials, seed yield is a crucial indicator in faba bean production, as it directly affects economic value and planting decisions [28]. In this study, we assessed the yield of 11 faba bean genotypes under both control and SA conditions. The yields decreased by 8.74% under SA treatment (Figure 1A), showing a statistically significant difference ($p < 0.05$). In the CK, the average yield was 3818.99 $t \cdot ha^{-1}$, while under SA treatment, the average yield decreased to 3368.81 $t \cdot ha^{-1}$. This finding differs from the results reported by Huang et al. and Teasley et al., who found that organic amendments and biochar can increase crop yields [29,30]. According to the Technical Specifications for Immobilizing Heavy Metals in Contaminated Soil [31], the yield reductions for genotypes F2, F4, and F11 exceeded the standard by 1.69, 1.17, and 1.24%, respectively, whereas reductions for other genotypes remained below the 10% threshold, within acceptable limits [31]. Although there were variations in yields among genotypes under the two treatments, these differences were not statistically significant. Further analysis revealed that the average yields of genotypes F1, F9, F3, and F6 were significantly higher than those of other genotypes under both treatment conditions, exceeding the average yields of all tested faba bean genotypes by 36.04, 29.68, 20.38, and 7.09, respectively, demonstrating their strong yield potential.

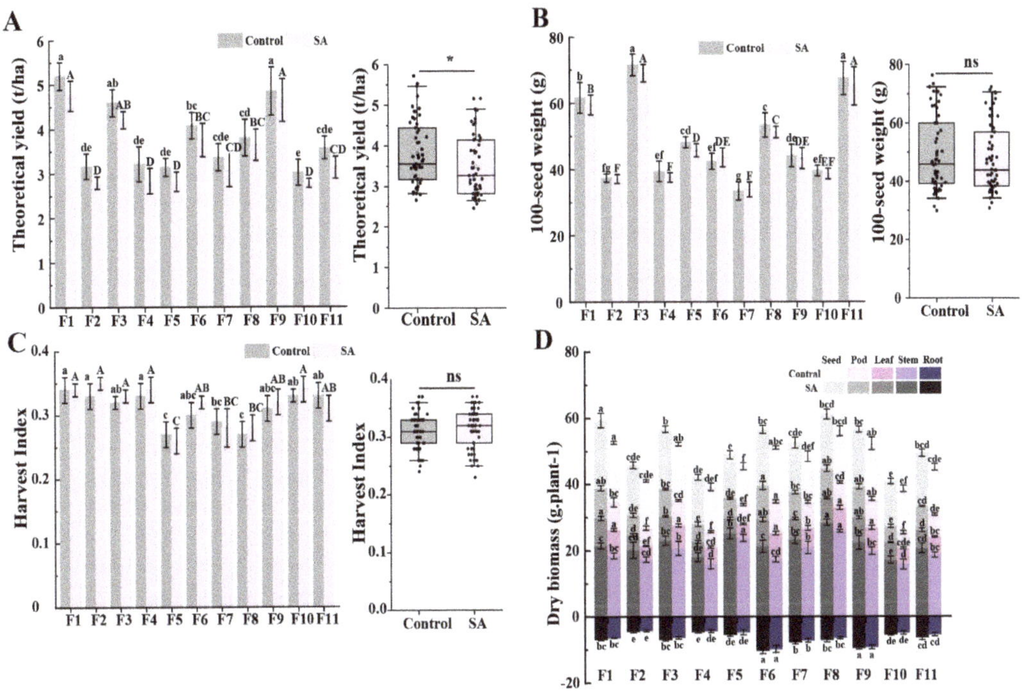

Figure 1. (**A**) Theoretical yield, (**B**) 100-seed weight, (**C**) harvest index (HI), and (**D**) dry biomass of 11 faba bean genotypes under control and SA treatments. Significant differences at $p < 0.05$ between groups are indicated by different letters. * represents a significant difference at $p < 0.05$ compared to the control, and "ns" denotes no significant difference.

Our findings provide practical guidance for faba bean production. Genotypes F1, F9, F3, and F6, which exhibited excellent yields, maintained high production rates even under SA treatment conditions, making them recommended varieties for farmers. These genotypes sustained high yields and demonstrated adaptability under SA treatment, making them valuable choices for optimizing faba bean planting yields and improving soil health management.

2.1.2. Harvest Index (HI) and Other Agronomic Traits Adjustment

No significant differences were observed in the 100-seed weight and HI of faba beans under SA treatment compared to those of CK (Figure 1B,C). Similarly, Figure S2 indicates that the plant height, number of seeds per pod, and SPAD values of mature leaves did not differ significantly between the two treatments, suggesting that these growth parameters were unaffected. Further analysis revealed that the 100-seed weight of genotypes F1, F3, and F11 was significantly higher, ranging from 59.50 to 71.51 g. Although the 100-seed weight of most genotypes was slightly reduced under SA treatment (all within 5%), the 100-seed weight of F6 and F2 increased by 2.78% (43.61 g) and 1.29% (37.19 g), respectively. This result suggests that F6 and F2 responded more favorably to SA treatment with respect to 100-seed weight, whereas F1, F3, and F11 maintained superior seed weights and economic value.

In contrast, although the HI of faba beans was insensitive to the SA treatment, dry biomass responded significantly ($p < 0.001$). Whole plant dry weights were reduced by 7.56% under SA treatment, with an average reduction of 7.43% in the shoot biomass. Figure S2F shows that the disease rate decreased by 45.33% after SA treatment. This result was verified by Renaud et al. and Tao et al., who reported that different compositions of soil amendments may affect soil anurans and soil microorganisms, thus affecting the health status of crops [32,33]. Based on our results, we reasonably assumed that the reduction in biomass by specific amendments may represent an adaptive response to the environment to reduce the incidence of pests and diseases; however, additional experiments are needed to confirm this speculation. Additionally, as shown in Figure S2B, faba bean stem dry biomass accounted for 34.04 to 52.04% of the total plant biomass across the 11 genotypes, suggesting that stems may represent an important site for mineral element accumulation.

2.2. Cd Cycling in Faba Bean Soil System in Response to Soil Amendment

2.2.1. Decreased Bioavailability of Cd in Soil

After SA treatment, the pH value increased from 5.39 ± 0.07 to 6.00 ± 0.06, reflecting an 11% increase (Table S1). Although the change between SA and CK was not statistically significant, it effectively mitigated soil acidification. By improving the soil pH, an SA treatment can mitigate the adverse effects of soil acidity on plant health and metal uptake rates. Yang et al. emphasized that strongly acidified and heavily contaminated rice soils are harmful to crop cultivation [17]. The present study demonstrated that an SA treatment had a substantial impact on improving soil properties and reducing Cd bioavailability.

The organic matter content increased from 20.98% to 23% following the SA treatment, with no statistically significant difference compared with CK. Studies by Buss et al. have shown that increased organic matter improves soil structure, water retention, and nutrient availability, and is closely correlated with soil acidity, microbial activity, and mineral absorption capacity [34–37]. A more pronounced effect was observed with regard to calcium (Ca) content; the change in total Ca content was statistically significant, showing a 59% increase while the exchangeable Ca content also increased significantly, with a 92% increase. Ji et al. and Wang et al. suggested that increased Ca content and bioavailability may alter the pH-DOC balance in the soil solution, affecting Cd uptake by crops [38,39]. In our study, increases in pH, organic matter, and Ca concentration, along with the enhanced exchangeable Ca, may have influenced the bioavailability of Cd in the soil.

Studies have shown that the concentration of DTPA-extractable Cd in soil is a key factor in determining Cd levels in plants. Exploring the effects of soil amendments on DTPA-Cd is crucial for enhancing food security [40,41]. In this study, the DTPA-extractable Cd (DTPA-Cd) concentrations decreased significantly by 37% following SA treatment (Table S1). Lower extractable Cd levels in soil are crucial for minimizing the risk of Cd uptake by plants, which can otherwise lead to contamination of the food chain and potential health risks for consumers. These findings suggest that an SA treatment can be a valuable strategy for improving soil health and nutrient availability. The significant improvements in pH and calcium levels, and the reduction in Cd bioavailability, highlight the potential benefits of SA for crop productivity and safety.

2.2.2. Reduced Cd Concentration in Faba Beans

Reducing the Cd content in the seeds was one of the key focuses of our study. As shown in Figure 2A, SA treatment reduced the average Cd content in seeds to 0.06 mg·kg^{-1}, a 60% decrease from 0.15 mg·kg^{-1} under CK ($p < 0.001$), allowing the seeds to meet food safety standards. Among the 11 faba bean genotypes, F1, F2, F3, F4, and F7 had Cd levels below 0.04 mg·kg^{-1}, whereas F10 contained 0.1 mg·kg^{-1}, and the remaining genotypes ranged between 0.06 and 0.08 mg·kg^{-1}. The reduction in Cd content varied from 33.33 to 81.25%, with F3 showing the highest reduction rate.

A

	Seed Cd		Pod Cd		Stem Cd		Leaf Cd	
	CK	SA	CK	SA	CK	SA	CK	SA
F3	0.16	**0.03** ***	0.68	0.16 ***	1.03	0.22 ***	1.13	0.74 **
F1	0.11	**0.03** ***	0.39	0.24 ***	0.62	0.19 ***	0.39	0.13
F7	0.13	**0.04** ***	0.49	0.18 ***	0.96	0.05 ***	1.44	0.13 ***
F4	0.2	**0.04** ***	0.75	0.35 ***	0.36	0.27	1.25	0.10 ***
F2	0.15	**0.04** ***	0.59	0.20 ***	0.66	0.13 ***	1.13	0.23 ***
F6	0.18	**0.06** ***	0.84	0.24 *	0.64	0.22 ***	1.39	0.12 ***
F5	0.17	**0.06** ***	0.59	0.17 ***	0.98	0.24 ***	1.28	0.19 ***
F9	0.12	**0.07** *	0.3	0.16	0.68	0.17 ***	1.01	0.12 ***
F11	0.14	**0.08** *	0.48	0.28 ***	1.15	0.36 ***	1.13	0.13 ***
F8	0.12	**0.08**	0.68	0.07 ***	0.57	0.21 ***	1.18	0.17 ***
F10	0.15	**0.10** *	0.74	0.41 ***	0.94	0.12 ***	0.96	0.20 ***
Avg.	0.15	**0.06** ***	0.59	0.22 ***	0.78	0.20 ***	1.12	0.21 ***

B

	TF (Seed/Stem)		TF (Seed/Pod)		TF (Seed/Leaf)		TF (Leaf/Stem)	
	CK	SA	CK	SA	CK	SA	CK	SA
F3	0.16	**0.15**	0.23	0.21	0.14	0.05	1.1	3.36 ***
F4	0.57	**0.16** ***	0.27	0.13	0.16	0.44	3.56	0.38 ***
F1	0.17	**0.19**	0.28	0.14	0.28	0.25	0.63	0.77
F11	0.12	**0.23**	0.28	0.28	0.12	0.63 ***	0.99	0.36
F5	0.17	**0.25**	0.28	0.36	0.13	0.32	1.33	0.80 *
F2	0.23	**0.28**	0.26	0.18	0.14	0.16	1.71	1.81 ***
F6	0.29	**0.28**	0.22	0.27	0.13	0.52 ***	2.17	0.57
F8	0.2	**0.40**	0.17	1.26 ***	0.1	0.54 ***	2.08	0.81 ***
F9	0.19	**0.41** *	0.42	0.44	0.12	0.55 ***	1.51	0.75 *
F7	0.14	**0.74** ***	0.26	0.23	0.09	0.31	1.5	2.44 **
F10	0.17	**0.78** ***	0.21	0.24	0.16	0.47 *	1.03	1.67
Avg.	0.22	**0.35** **	0.26	0.34	0.14	0.39 ***	1.6	1.25

C

	BCF (Seed/Soil)		BCF (Pod/Soil)		BCF (Stem/Soil)		BCF (Leaf/Soil)	
	CK	SA	CK	SA	CK	SA	CK	SA
F3	0.15	**0.03** ***	0.66	0.16 ***	0.99	0.22 ***	1.09	0.73 **
F1	0.1	**0.03** ***	0.38	0.24	0.6	0.18 ***	0.37	0.13
F2	0.15	**0.04** ***	0.57	0.20 ***	0.64	0.13 ***	1.08	0.23 ***
F4	0.19	**0.04** ***	0.72	0.35 ***	0.34	0.27	1.2	0.1
F7	0.12	**0.04** ***	0.47	0.18 ***	0.92	0.05 ***	1.38	0.13 ***
F5	0.16	**0.06** ***	0.57	0.17 ***	0.94	0.24 ***	1.23	0.19 ***
F6	0.18	**0.06** ***	0.81	0.24 ***	0.62	0.22 ***	1.34	0.12 ***
F9	0.12	**0.07** *	0.29	0.16	0.65	0.16 ***	0.97	0.12 ***
F8	0.11	**0.08**	0.65	0.07 ***	0.55	0.21 ***	1.14	0.17 ***
F11	0.13	**0.08**	0.47	0.28 ***	1.11	0.36 ***	1.09	0.13 ***
F10	0.15	**0.10**	0.71	0.41 ***	0.91	0.12 ***	0.92	0.20 ***
Avg.	0.14	**0.06** ***	0.57	0.22 ***	0.75	0.20 ***	1.07	0.20 ***

Figure 2. (**A**) Cadmium (Cd) concentrations in the aerial tissues of faba beans, including seed, pod, stem and leaf; (**B**) translocation factor (TF) and (**C**) BCFs of Cd in the aerial parts of faba beans. The significance between SA and CK treatments is as follows: *** for $p < 0.001$, ** for $p < 0.01$, * for $p < 0.05$, and no marking indicates no significant difference. The light purple color indicates the corresponding smaller value and the dark purple color indicates the corresponding larger value.

Similarly, SA treatment had a profound impact on Cd levels in shoot tissues compared with CK. SA treatment significantly lowered Cd levels in the pods, stems, and leaves (Figure 2A) ($p < 0.001$). Compared to CK, the Cd content decreased to 62.71, 74.36, and 81.25% under SA treatment, respectively. Given that stems constitute the largest plant biomass, their Cd levels were crucial for assessing the overall accumulation. The reduction percentages for stem Cd were highest in F7 (94.79%) and lowest in F4 (25.00%), indicating that most genotypes responded sensitively to SA treatment in terms of stem Cd accumulation, making them suitable for selection in SA-treated Cd-contaminated soils.

This study demonstrates that SA treatments significantly reduce the Cd content in the seeds, pods, stems, and leaves of faba beans, thereby enhancing crop safety in Cd-Cd-contaminated soils. The faba bean genotypes F3, F1, F2, F4, and F7 exhibited low Cd content in seeds and showed significant Cd reductions in response to SA treatment, making them excellent candidates for safe crop selection in moderately to highly Cd-contaminated fields.

2.2.3. TFs and BCFs of Cd in Faba Beans

An assessment of TFs and BCFs can provide a comprehensive understanding of Cd dynamics in faba beans under SA treatment. As shown in Figure 2B, TF (seed/stem) increased from 0.22 to 0.35 ($p < 0.01$), with F7, F10, and F9 showing significant increases (0.74, 0.78, and 0.41, respectively) compared to CK. In contrast, the TF (seed/stem) of F4 decreased to 0.16, indicating an altered Cd translocation efficiency in certain genotypes. The TF (seed/leaf) also increased by 178.6%, from 0.14 to 0.39, with significant improvements in genotypes F6, F8, F9, F11, and F10, suggesting enhanced Cd translocation from leaves to seeds. However, the TF (seed/pod) increased marginally from 0.26 to 0.34 without statistical significance for SA, and TF (leaf/stem) showed variable responses among genotypes, with notable increases in F2 and F7 and decreases in F4, F6, F8, and F9 in SA. The observed increases in TFs, particularly TF (seed/stem) and TF (seed/leaf), highlight the differential Cd transport capabilities among the various genotypes. The BCF of the shoots reduced the Cd content, indicating that SA treatment effectively decreased Cd absorption and accumulation in faba beans ($p < 0.001$) (Figure 2C).

Based on the TF (seed/stem), genotypes F7, F10, and F9 were identified as optimal for Cd remediation because of their high translocation efficiency from the stem to the seed. This result may be attributed to genetic characteristics or physiological responses specific to different genotypes. The capacity for Cd mobilization from the leaves to the seeds was significantly enhanced in genotypes F6, F8, F9, F11, and F10, indicating that these genotypes were better suited for Cd accumulation. The combination of high TFs and reduced BCFs indicated that SA treatment improved the Cd translocation efficiency and underscored the genetic or physiological adaptability of different genotypes in Cd-contaminated farmland.

2.2.4. Linear Correlation of Cd Concentrations in Faba Bean Tissues

To further investigate the distribution and translocation patterns of Cd in different faba bean tissues, we conducted a correlation analysis of Cd concentrations in the seeds, pods, stems, and leaves of 11 faba bean genotypes under CK and SA conditions (Figure 3). The results revealed highly significant linear correlations ($p < 0.001$) between the Cd concentrations in the edible parts of faba beans and those in the pods, stems, and leaves. Specifically, the correlation coefficients R^2 between the Cd concentration in the seed and the Cd concentrations in the seed, pod stem, and leaf were 0.70371, 0.59545, and 0.48515, respectively. The highest regression coefficient (0.19379), which was observed for the pod, indicates the strongest linear correlation of the Cd concentration between seed and pod. This result suggests that Cd accumulation in the pods most significantly affected the seed

Cd concentration, followed by the leaves and stems. Consequently, Cd accumulation in faba bean seeds was influenced by multiple tissues rather than an individual tissue. Thus, comprehensively controlling Cd accumulation across different tissues may more effectively reduce seed Cd concentrations and improve crop safety.

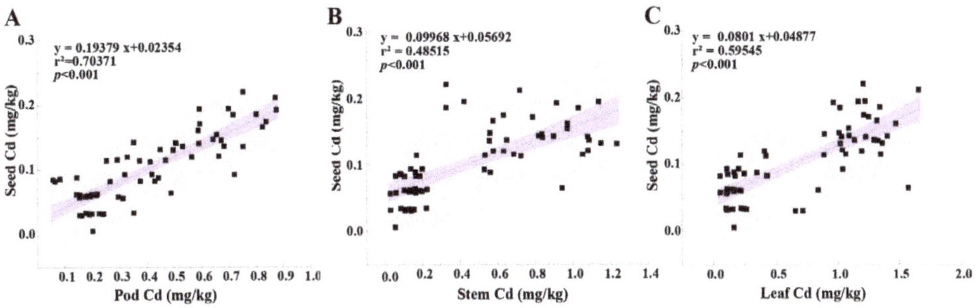

Figure 3. Linear correlation analysis of Cd concentrations in faba bean tissues. The solid line represents the best-fit linear regression and the shaded area indicates the 95% confidence interval. Sample sizes for (**A–C**) are N = 66 each.

2.2.5. Shoot Cd Uptake in Faba Beans

Figure 1D clearly illustrates the dry biomass of various aboveground tissues of the faba bean genotypes. By integrating the results from Figure 2A, we accurately calculated the Cd uptake for each genotype under SA treatment and control conditions using the total accumulation of the Cd analysis formula (Figure 4A) and precisely determined the proportion of Cd uptake in each tissue (Figure 4B). Figure 4A demonstrates that all tested faba bean genotypes exhibited significant differences in total Cd uptake in shoots under SA treatment compared to the control ($p < 0.001$ for all genotypes). This result suggests that SA treatment significantly affected Cd uptake and distribution in faba beans and potentially played a crucial role in regulating Cd transport mechanisms. Specifically, under control conditions, the total shoot Cd uptake across all genotypes ranged from 2.54 to 4.96 $g{\cdot}ha^{-1}$ (with an average of 4.09 $g{\cdot}ha^{-1}$). However, under SA treatment, the total Cd uptake was reduced by 75%, with a range of 0.63 to 1.51 $g{\cdot}ha^{-1}$ (average of 1.10 $g{\cdot}ha^{-1}$). The genotypes F7, F3, and F6 showed the most pronounced reductions, with decreases of 90, 80, and 79%, respectively.

Figure 4B shows the Cd uptake distribution across different shoot tissues in the 11 faba bean genotypes under CK and SA treatment. In CK, stems were the main site of Cd uptake, accounting for 34% to 68% of the uptake (with an average of 55%). Under SA treatment, stems remained the primary site, with an average uptake of 50%, representing an 5% decrease compared to their uptake when receiving CK treatment. However, there were notable differences in the different genotypes' responses to SA treatment. Specifically, genotypes F7, F10, F3, and F1 exhibited a significant reduction in the percentage of Cd in the stem, whereas F4 and F8 showed a significant increase, and the other genotypes showed no significant change. This result indicates that although SA treatment effectively reduced the overall Cd uptake and accumulation by modulating Cd transportation, the distribution patterns of Cd within faba beans remained generally stable, with significant variations in Cd distribution mechanisms among different genotypes.

SA treatment significantly reduced Cd uptake in the shoots of faba beans, with genotypes F7, F3, and F6 showing the most pronounced responses, and affected the tissue-specific distribution of Cd across the different genotypes (Figure 4). For all tested faba bean genotypes, regardless of whether CK or SA treatment was administered, the stem was generally the primary site of Cd accumulation. However, in certain genotypes, the proportion of Cd in the stem varied significantly under SA treatment. This result suggests that although SA treatment reduced shoot Cd accumulations by modulating Cd transport mechanisms, its impact on Cd distribution differed among genotypes. These findings suggest that SA treatment can potentially improve the management of Cd content in faba beans by optimizing Cd transport and distribution mechanisms, as well as provide insights into optimizing Cd management strategies tailored to specific genotypes. This finding has important implications for the breeding of faba bean genotypes with enhanced Cd tolerance and regulatory capacity.

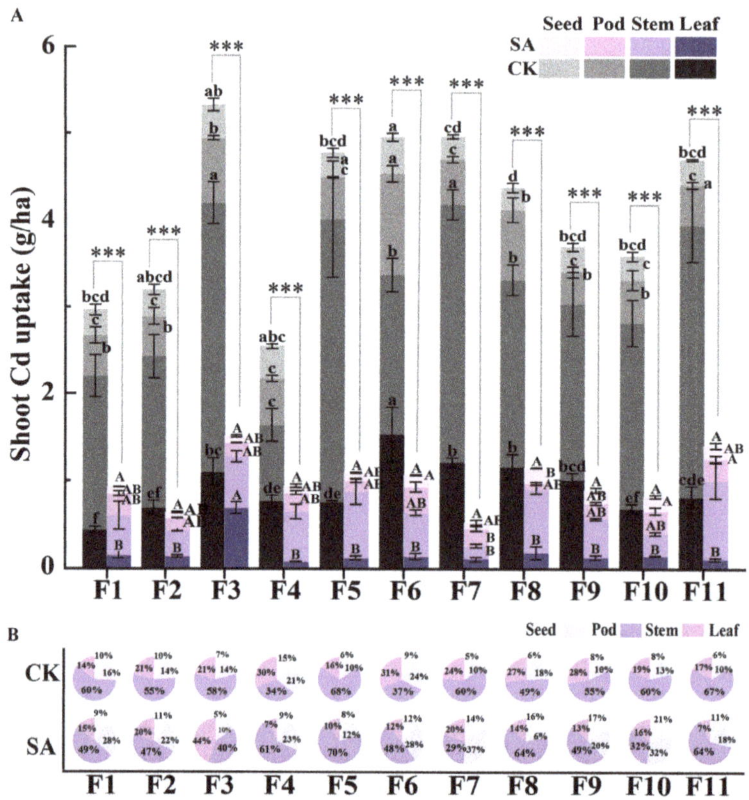

Figure 4. Comparison of Cd uptake and distribution in the shoots of 11 faba bean genotypes under SA and CK treatments. (**A**) Shoot Cd uptake (g·ha^{-1}) and its significant differences. (**B**) Percentage of cadmium uptake in four tissues (seed, pod, stem, leaf) within the shoots. Groups marked with different letters exhibit a significant difference at $p < 0.05$. *** indicate a significant difference ($p < 0.001$) when compared to CK.

2.3. *Absorption and Distribution of Mineral Elements in Faba Bean Seeds by Soil Amendment*

2.3.1. Correlation Between Elements and Yield in Seeds

In this study, SA treatment clearly and significantly reduced the Cd concentration in faba bean seeds; however, the yield was also slightly reduced. Therefore, we hypothesize that SA treatment may affect the uptake and accumulation of other elements in faba bean seeds, thereby affecting yields. As shown in Figure 5A and Figure S3, CK and SA were divided into two clusters, which reflects the unique role of SA treatment in regulating element accumulation and crop performance. In addition, SA treatment positively affected the accumulations of Ca, S, Mo, P, Mg, and Cu, whereas it had no effect or even an inhibitory effect on the accumulations of Fe, Zn, Co, Al, Cd, and Mn. The six elements Ca, S, Mo, P, Mg, and Cu showed similar behavioral patterns in response to SA treatments, which may indicate their synergistic effects or interdependence in uptake, transportation, or metabolism in plants. Trace elements such as Fe, Zn, Co, Al, Cd, and Mn were clustered with yields, implying that the accumulation levels of these elements may have jointly influenced the key physiological processes of faba bean growth and yields and that changes in these elements might cause major changes in yields under an SA treatment.

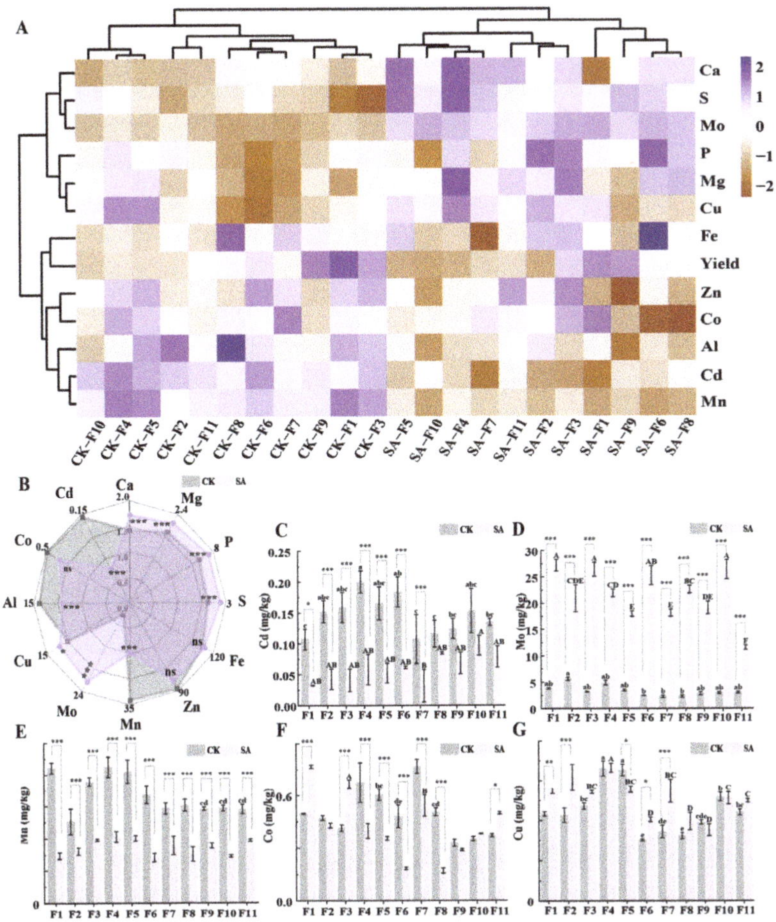

Figure 5. Correlation analysis between elemental contents and yield, and modulations in elemental concentrations in faba bean seeds after SA treatment. (**A**) Clustered heatmap of seed metal elements and yield in CK and SA treatment for all tested faba bean genotypes. (**B**) Radar chart comparing all elements' concentrations after SA treatment vs. CK, with significance marked within the purple area. (**C–G**) Concentrations of the trace elements Cd, Mo, Mn, Al, and Cu in seeds, each represented individually. Gray represents CK, and pink and purple represent SA. Ca, Mg, P, and S are in g·kg^{-1}; Fe, Zn, Mn, Mo, Cu, Al, Ni, Co, and Cd are in mg·kg^{-1}. Significant differences at $p < 0.05$ between groups are indicated by different letters. *, **, *** represent, respectively, significant differences at $p < 0.05$, $p < 0.01$, and $p < 0.001$ compared to CK, while non-marked differences represent no significance.

2.3.2. Modulation of Elemental Concentrations in Seeds

Figure 5B shows significant increases in Ca, Mg, P, S, Mo, and Cu in seeds under SA treatment, whereas Al, Cd, and Mn decreased significantly. The Fe, Zn, and Co levels remained unchanged. Specifically, the macroelements Ca, Mg, P, and S increased by 21.26%, 13.50%, 12.94%, and 15.77%, respectively, while Mo and Cu increased by 544.42% and 13.62%. Cd, Mn, and Al exhibited reductions of 59.63%, 48.70%, and 24.68%, respectively. Mn decreased by 47.54% in line with Cd reduction. Mo content increased between 271.26% and 918.09%, indicating enhanced Mo uptake and translocation. Cu content varied among genotypes, with decreases in F5, F9, and F10, whereas Fe, Zn, and Co responses were genotype-specific. Notably, Zn increased in F11, F2, and F3, whereas Co increased in F3, F1, and F11. Fe contents increased significantly in F6, F5, F3, and F2.

Overall, SA treatment effectively improved the nutrient content of faba bean seeds through genotype-specific responses. F11 combined with SA is suitable for environments that require high Zn and Co concentrations, whereas F6 combined with SA is ideal for iron-deficient areas. Monitoring Mo levels is crucial to prevent interference with the absorption of other nutrients.

2.4. Nutritional Quality and Mineral Bioavailability in Seeds

SA treatment exhibited significant effects on mineral regulation but also led to a slight reduction in seed yield. The decrease in yield suggests that SA treatment may influence other key seed components such as proteins and phytic acid. Proteins, as a core nutritional component of faba beans, directly affect their nutritional value, whereas phytic acid, as a major form of phosphorus storage and an anti-nutritional factor, affects mineral absorption and utilization. Given that SA treatment significantly increased the levels of P and S in seeds, it was crucial to assess changes in phytic acid and protein content to comprehensively evaluate the impact of SA treatment.

Despite the significant increase in P and S levels, the protein and phytic acid content did not exhibit significant changes across all genotypes under SA treatment (Figure 6A). This indicates that SA treatment primarily affected the accumulation of specific minerals rather than the synthesis or accumulation of proteins and phytic acid, possibly because of the inherent stability of the biosynthetic pathways for these components. This stability helps maintain the nutritional quality of the seeds, even with variations in mineral content. However, for specific genotypes, such as F5 and F11, notable changes were observed: phytic acid content in the F5 genotype decreased by 64% and protein content in the F11 genotype decreased by 21% under SA treatment, which may be attributed to the genotype-specific responses of faba beans.

The ratios of phytic acid to minerals were analyzed to further understand the effect of SA treatment on nutritional quality. Under SA treatment, the Phy/Ca (Figure 6B,C), Phy/Mg (Figure S4A,D), and Phy/P (Figure 6D,E) ratios in faba bean seeds were significantly reduced by approximately 23, 20, and 19%, respectively (all $p < 0.05$). However, compared with the CK treatment, the Phy/Fe (Figure S4B,E) and Phy/Zn (Figure S4C,F) ratios showed no significant differences under SA treatment, with a 3% increase and an 8% decrease, respectively. As shown in Figure 6C,D, SA treatment significantly reduced the Phy/Ca ratio in the F5 genotype by 76%, and the Phy/P ratios in the F6, F5, and F8 genotypes by 71%, 61%, and 48%, respectively. Additionally, Figure S4D shows reductions of 65% and 60% in the Phy/Mg ratios for the F6 and F5 genotypes under SA treatment. These findings suggest that SA treatment significantly reduced the phytic acid–mineral ratios (e.g., calcium, magnesium, and phosphorus) in specific faba bean genotypes like F5 and F6, potentially enhancing mineral bioavailability. This indicates that targeted SA application could improve mineral absorption efficiency without notably affecting protein or phytic acid content, enhancing the nutritional value of faba beans. These insights open up new possibilities for developing nutritionally improved faba bean varieties and present promising directions for crop improvement and agricultural production.

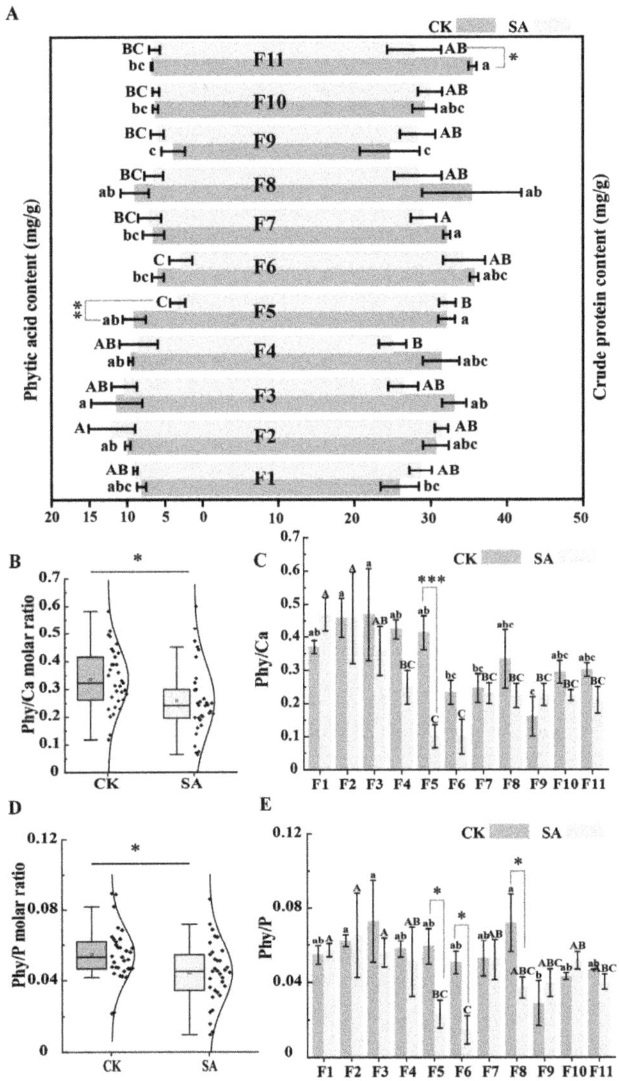

Figure 6. Protein content, phytic acid content, and mineral bioavailability (phytic acid/mineral ratio) in faba bean genotypes. (**A**) Crude protein content (mg·g^{-1}) and phytic acid content (mg·g^{-1}) in 11 faba bean genotypes under SA treatment compared to CK. (**B**,**C**) depict the response of the Phy/Ca molar ratio in all faba bean genotypes to SA treatment. (**D**,**E**) illustrates the response of the Phy/P molar ratio under the same SA treatment. Different letters signify significant differences between groups at $p < 0.05$. * indicates a significant difference at $p < 0.05$, ** at $p < 0.01$, *** at $p < 0.001$, and non-marked differences represent no significance.

3. Materials and Methods

3.1. Materials and Field Experiments

We conducted a field experiment in moderately Cd-contaminated acidic soil to evaluate the performance of 11 faba bean genotypes treated with SA. The field experiment was conducted in Cd-contaminated agricultural land in Lijin Garden Village, Quzhou City, Zhejiang Province. The study employed a randomized complete block design with three replicates to ensure robust comparisons. The experiment included both a control group and a treatment group. The treatment group was supplied with a soil amendment (SA), Greenlife No. 1 [42] (Purchased from Greenlife Technologies, Inc., Hangzhou, China), at a rate of 6000 kg·ha^{-1} alongside basic fertilizers, while the control group received only basic fertilizers without any soil amendment. This approach allowed us to comprehensively assess the effects of soil amendment on crop performance under moderately Cd-contaminated conditions. All tested faba bean genotypes were sourced from local seed companies in Zhejiang Province, China, and are detailed in Table S2 and Figure S1. The genotypes, referred to as F1, F2, . . ., F11, were obtained from markets in Hangzhou and are commonly used by Chinese farmers. Specifically, we examined changes in soil properties, Cd bioavailability, crop yields, and concentrations of essential macronutrients and micronutrients in faba bean seeds. This study aimed to evaluate the effects of soil amendment supply on faba bean cultivation in moderately Cd-polluted soil. Further details regarding the materials and field experiments are provided in Text S1.

3.2. Plant Samples

3.2.1. Yield and Biomass Analysis

All genotypes of faba beans were grown for 220 days in moderately Cd-contaminated farmland, and after harvest, estimates of theoretical seed yield and disease rates were calculated. Following classification and analysis, the dry weights of the plants were measured, and relevant agronomic indices and mineral content of different tissues, as well as the nutritional composition of seeds, were determined. The specific experimental procedures for analyzing the above indicators in the laboratory are detailed in Text S2.

3.2.2. Metal Element Analysis

The acid digestion method was used to digest all plant tissues. Macroelements were analyzed by ICP-OES and microelements by ICP-MS. The digestion and determination processes for the metal elements are detailed in Text S3.

3.2.3. Analyses of Protein and Phytic Acid Contents and Phy/M

The crude protein content in the faba bean seeds was determined via the classical Kjeldahl nitrogen determination method to obtain the nitrogen content of the seeds [43]. The crude protein content was calculated by multiplying the nitrogen content by a conversion factor of 6.25 [44]. This method ensures an efficient and accurate determination of crude protein content in seed samples.

The phytic acid content of faba bean seeds was determined using the ferric chloride colorimetric method and the absorbance value of the solution to be tested was measured at 500 nm using an enzyme marker [43,45]. The specific operation and measurement process can be found in Text S4.

The formulas for phytate–mineral molar ratios (Phy/M) in the seed of faba bean plants are as follows:

$$\mathrm{Phy/M} = \left(C_{Phy}/660\right)/(C_M/M_r)$$

C_{Phy} (mg·kg^{-1}) and C_M (mg·kg^{-1}) denote the contents of phytic acid and mineral elements in faba bean seeds; M_r indicates the mineral element.

3.2.4. Analysis of Bioaccumulation Factor, Translocation Factor, and Total Cadmium Accumulation

The formulas used to calculate the bioaccumulation factor (BCF), translocation factor (TF), and total accumulation of Cd (TA$_{\text{Cd}}$, g·ha^{-1}) in the aboveground parts of the faba bean plants were as follows:

$$\text{BCF} = C_{up}/C_{soil}$$

$$\text{TF} = C_{up}/C_{low}$$

$$\text{TA}_{\text{Cd}} = \sum Cs_{Cd} \times Biomass_n$$

C_{up}, C_{low}, and C_{soil} denote the Cd concentrations in the upper and lower tissues of faba bean, as well as in the soil, respectively. Cs_{Cd} indicates the concentration of Cd in the shoots of faba beans and $Biomass_n$ refers to the biomass (g·ha^{-1}) across different aboveground faba bean tissues.

3.3. Soil Sample Collection and Physicochemical Property Analysis

Soil samples were collected at a depth of 10–20 cm from three areas in the CK and SA groups using a stainless-steel shovel. After sampling using the five-point plum throat method, the samples were mixed, air-dried, ground, and sieved through a 100-mesh screen, resulting in three pre-planting soil samples and six post-harvest faba bean soil samples (three from CK and three from SA) that were stored in sealed plastic bags [43]. The moderately Cd-contaminated soil was a brown loam, and the basic properties of the soil were determined following the risk control standards for environmental quality of contaminated soil in agricultural soil in China, as shown in Table S3. The specific determination and analysis of soil collection and the soil's physicochemical properties are described in Text S5.

4. Statistical Analysis

All data are presented as the mean ± standard deviation (SD) of the three replicates. Statistical analysis was performed using the IBM SPSS Statistics software (version 26.0; one-way ANOVA followed by Tukey's post hoc test (with a significance level of $p < 0.05$) was used to determine significant differences between treatment groups. Data were visualized using the Origin 2023 software. In all figures and tables, unless otherwise specified, data are presented as a mean ± SD, with three replicates per treatment, and different uppercase and lowercase letters indicate significant differences between faba bean genotypes ($p < 0.05$). The significance levels between SA and CK treatments are denoted as follows: *** for $p < 0.001$, ** for $p < 0.01$, * for $p < 0.05$, while "ns" and non-marked differences in figures indicate no significant difference.

5. Conclusions

In this study, we used the strategy of soil amendment (SA) intercropping with a Cd-low-accumulating crop, faba bean (*Vicia faba* L.), to address food security issues in Cd-contaminated farmland. The application of SA was effective in reducing the bioavailability of Cd, increasing the acidity of the soil, and significantly reducing Cd accumulation in faba bean seeds, ensuring that most of the tested genotypes of faba beans met national food security standards. The application of SA significantly increased the essential mineral nutrients (Ca, Mg, P, and S) and their bioavailability, increased the micronutrients Mo and Cu, and reduced Mn and Al accumulations. There was a slight decrease in faba bean yields, which remained within the acceptable range of national standards. In this study, three faba bean genotypes, F3, F5, and F6, had the most balanced combined performance in terms of nutrient quality, food security, and yield in moderately Cd-contaminated soils in synergy with SA. These findings provide valuable insights into mitigating Cd contamination in faba beans using SA, thereby contributing to safer food production and improved nutrient composition in Cd-contaminated farmland. However, further research is needed to confirm these findings across various soils and environments, assess long-term effects on crop yields and quality, and explore interactions with other soil amendments. Overall, this study offers a valuable foundation for developing sustainable strategies to manage cadmium contamination and enhance nutrient-rich crop production.

Supplementary Materials: The following supporting information can be downloaded at https://www.mdpi.com/article/10.3390/plants14010141/s1, Figure S1: The weather conditions throughout the entire growth period of faba bean. Figure S2: Plant height, ground biomass, number of pods per plant, number of seeds per pods, SPAD values, and disease score of 11 fava bean genotypes under control and SA treatments. Figure S3: Modulations in elemental concentrations in fava bean seeds after SA treatment vs. CK; Figure S4: Phytic acid ratios with Mg, Zn, and Fe. Table S1: The main physicochemical properties of soil under CK and SA treatments; soils were sampled after the faba beans were harvested. Table S2: Basic traits of 11 faba bean genotypes. Table S3: The fundamental physicochemical characteristics of the tested soil. Text S1: Basic traits of 11 faba bean genotypes and details of field experiments. Text S2: Details of yield and biomass analysis of plant samples. Text S3: Determination and analysis of elements in plant samples. Text S4: Determination and analysis of phytic acid content. Text S5: Determination and analysis of soil collection and the soil's physicochemical properties.

Author Contributions: L.C.: Data curation, Formal analysis, Investigation, Methodology, Validation, Writing—original draft, Writing—review and editing. J.L.: Conceptualization, Methodology, Data curation. X.W.: Visualization, Investigation, Methodology. M.A.M.M.: Formal analysis, Writing—review and editing. X.H.: Methodology, Visualization. Z.H.: Formal analysis, Writing—review & editing, Supervision. C.X.: Investigation, Resources. W.T.: Project administration, Resources, Supervision, Writing—review and editing. X.Y.: Funding acquisition, Project administration, Resources, Supervision, Writing—review and editing. All authors have read and agreed to the published version of the manuscript.

Funding: Funding for this project was provided by the Department of Science and Technology of Zhejiang Province (#2023C02002), and the Key Project of Zhejiang Agricultural Bureau (#2020SNLF009).

Data Availability Statement: This manuscript does not report data generation or analysis. Data are available from the authors upon request.

Acknowledgments: The authors thank Hongfei Zhang (Zhejiang University) for the assistance with the ICP-OES/MS measurement.

Conflicts of Interest: The authors declare that they have no known competing financial interests or personal relationships that could have appeared to influence the work reported in this paper.

Abbreviations

SA—soil amendment; BCF—bioaccumulation factor; TF—translocation factor; HI—harvest index.

References

1. Globocan 2018 Latest Global Cancer Data. Available online: https://www.iarc.who.int/infographics/globocan-2018-latest-global-cancer-data (accessed on 2 January 2025).
2. Bai, B.; Kong, S.; Root, R.A.; Liu, R.; Wei, X.; Cai, D.; Chen, Y.; Chen, J.; Yi, Z.; Chorover, J. Release Mechanism and Interactions of Cadmium and Arsenic Co-Contaminated Ferrihydrite by Simulated In-Vitro Digestion Assays. *J. Hazard. Mater.* **2024**, *467*, 133633. [CrossRef] [PubMed]

3. Djordjevic, V.R.; Wallace, D.R.; Schweitzer, A.; Boricic, N.; Knezevic, D.; Matic, S.; Grubor, N.; Kerkez, M.; Radenkovic, D.; Bulat, Z.; et al. Environmental Cadmium Exposure and Pancreatic Cancer: Evidence from Case Control, Animal and In Vitro Studies. *Environ. Int.* **2019**, *128*, 353–361. [CrossRef] [PubMed]
4. Lin, X.-Y.; Ye, T.-W.; Duan, X.; Wang, B.-X.; Zhou, D.; Li, H.-B. Cadmium in Market Pork Kidneys: A Study on Cadmium Bioavailability and the Health Effects Based on Mouse Models. *Environ. Sci. Technol.* **2024**, *58*, 14651–14661. [CrossRef]
5. Wang, Y.; Wang, D.; Hao, H.; Cui, J.; Huang, L.; Liang, Q. The Association between Cadmium Exposure and the Risk of Chronic Obstructive Pulmonary Disease: A Systematic Review and Meta-Analysis. *J. Hazard. Mater.* **2024**, *469*, 133828. [CrossRef]
6. China Bulletin of the National Soil Pollution Survey. Available online: https://www.gov.cn/foot/2014-04/17/content_2661768.htm (accessed on 7 September 2024).
7. Hussain, B.; Ashraf, M.N.; Shafeeq-ur-Rahman; Abbas, A.; Li, J.; Farooq, M. Cadmium Stress in Paddy Fields: Effects of Soil Conditions and Remediation Strategies. *Sci. Total Environ.* **2021**, *754*, 142188. [CrossRef]
8. Yang, Y.; Gu, Y.; Zhang, Y.; Zhou, Q.; Zhang, S.; Wang, P.; Yao, Y. Spatial–Temporal Mapping of Urine Cadmium Levels in China during 1980–2040: Dietary Improvements Lower Exposure amid Rising Pollution. *J. Hazard. Mater.* **2024**, *473*, 134693. [CrossRef]
9. EFSA Science, Safe Food, Sustainability. Available online: https://www.efsa.europa.eu/en (accessed on 7 September 2024).
10. Adhikari, S.; Moon, E.; Timms, W. Identifying Biochar Production Variables to Maximise Exchangeable Cations and Increase Nutrient Availability in Soils. *J. Clean. Prod.* **2024**, *446*, 141454. [CrossRef]
11. Lian, J.; Cheng, L.; Zhai, X.; Wu, R.; Huang, X.; Chen, D.; Pan, J.; Shohag, M.J.I.; Xin, X.; Ren, X.; et al. Zinc Glycerolate (Glyzinc): A Novel Foliar Fertilizer for Zinc Biofortification and Cadmium Reduction in Wheat (*Triticum aestivum* L.). *Food Chem.* **2023**, *402*, 134290. [CrossRef]
12. Vinson, V. Cooperating to Control Cadmium. *Science* **2019**, *366*, 321–322. [CrossRef]
13. Wu, Y.; Ma, L.; Liu, Q.; Vestergård, M.; Topalovic, O.; Wang, Q.; Zhou, Q.; Huang, L.; Yang, X.; Feng, Y. The Plant-Growth Promoting Bacteria Promote Cadmium Uptake by Inducing a Hormonal Crosstalk and Lateral Root Formation in a Hyperaccumulator Plant *Sedum alfredii*. *J. Hazard. Mater.* **2020**, *395*, 122661. [CrossRef]
14. Xiao, Q.; Zhang, W.; Wu, L.; Huang, Y.; Cai, Z.; Li, D.; Xu, X.; Hartley, I.P. Long-Term Liming Mitigates the Positive Responses of Soil Carbon Mineralization to Warming and Labile Carbon Input. *J. Environ. Manag.* **2024**, *354*, 120498. [CrossRef] [PubMed]
15. Zhu, Y.; Zhang, Q.; Li, Y.; Pan, Z.; Liu, C.; Lin, D.; Gao, J.; Tang, Z.; Li, Z.; Wang, R.; et al. Role of Soil and Foliar-Applied Carbon Dots in Plant Iron Biofortification and Cadmium Mitigation by Triggering Opposite Iron Signaling in Roots. *Small* **2023**, *19*, 2301137. [CrossRef] [PubMed]
16. Shi, L.; Guo, Z.; Liu, S.; Xiao, X.; Peng, C.; Feng, W.; Ran, H.; Zeng, P. Effects of Combined Soil Amendments on Cd Accumulation, Translocation and Food Safety in Rice: A Field Study in Southern China. *Environ. Geochem. Health* **2022**, *44*, 2451–2463. [CrossRef] [PubMed]
17. Yang, Y.; Wang, T.; Li, Y.; Wang, M.; Chen, W.; Dai, Y. Mitigating Cadmium Contamination of Rice Soils Supporting Tobacco–Rice Rotation in Southern China: Win–Win or Lose–Lose? *J. Hazard. Mater.* **2022**, *425*, 128052. [CrossRef]
18. Wang, Y.; Tang, D.; Yuan, X.; Uchimiya, M.; Li, J.; Li, Z.; Luo, Z.; Xu, Z.; Sun, S. Effect of Amendments on Soil Cd Sorption and Trophic Transfer of Cd and Mineral Nutrition along the Food Chain. *Ecotoxicol. Environ. Saf.* **2020**, *189*, 110045. [CrossRef]
19. Björnsdotter, E.; Nadzieja, M.; Chang, W.; Escobar-Herrera, L.; Mancinotti, D.; Angra, D.; Xia, X.; Tacke, R.; Khazaei, H.; Crocoll, C.; et al. VC1 Catalyses a Key Step in the Biosynthesis of Vicine in Faba Bean. *Nat. Plants* **2021**, *7*, 923–931. [CrossRef]
20. FAOSTAT. Available online: https://www.fao.org/faostat/en/#home (accessed on 8 September 2024).
21. Feedipedia: Animal Feed Resources Information System. Available online: https://www.feedipedia.org/node/4926 (accessed on 14 September 2024).
22. Xiao, Z.; Zou, D.; Zeng, X.; Zhang, L.; Liu, F.; Wang, A.; Zeng, Q.; Zhang, G.; Li, L. Cadmium Accumulation in Oilseed Rape Is Promoted by Intercropping with Faba Bean and Ryegrass. *Ecotoxicol. Environ. Saf.* **2020**, *205*, 111162. [CrossRef]
23. Ji, Y.; Liu, R.; Xiao, Y.; Cui, Y.; Chen, Z.; Zong, X.; Yang, T. Faba Bean Above-Ground Biomass and Bean Yield Estimation Based on Consumer-Grade Unmanned Aerial Vehicle RGB Images and Ensemble Learning. *Precis. Agric.* **2023**, *24*, 1439–1460. [CrossRef]
24. Krenz, L.M.M.; Grebenteuch, S.; Zocher, K.; Rohn, S.; Pleissner, D. Valorization of Faba Bean (*Vicia faba*) by-Products. *Biomass Convers. Biorefin.* **2023**, *14*, 26663–26680. [CrossRef]
25. Lin, Q.; Hamid, Y.; Yin, X.; Hussain, B.; He, Z.; Yang, X. Screening of Low-Cd Accumulating Early Rice Cultivars Coupled with Phytoremediation and Agro-Production: Bioavailability and Bioaccessibility Tests. *Sci. Total Environ.* **2022**, *844*, 157143. [CrossRef]
26. Dhull, S.B.; Kidwai, M.K.; Noor, R.; Chawla, P.; Rose, P.K. A Review of Nutritional Profile and Processing of Faba Bean (*Vicia faba* L.). *Legume Sci.* **2022**, *4*, e129. [CrossRef]
27. Segers, A.; Dumoulin, L.; Megido, R.C.; Jacquet, N.; Cartrysse, C.; Kamba, P.M.; Pierreux, J.; Richel, A.; Blecker, C.; Francis, F. Varietal and Environmental Effects on the Production of Faba Bean (*Vicia faba* L.) Seeds for the Food Industry by Confrontation of

Agricultural and Nutritional Traits with Resistance against *Bruchus* spp. (Coleoptera: Chrysomelidae, Bruchinae). *Agric. Ecosyst. Environ.* **2022**, *327*, 107831. [CrossRef]
28. Zhang, S.; Zhu, Q.; de Vries, W.; Ros, G.H.; Chen, X.; Muneer, M.A.; Zhang, F.; Wu, L. Effects of Soil Amendments on Soil Acidity and Crop Yields in Acidic Soils: A World-Wide Meta-Analysis. *J. Environ. Manag.* **2023**, *345*, 118531. [CrossRef] [PubMed]
29. Huang, Y.; Tao, B.; Lal, R.; Lorenz, K.; Jacinthe, P.-A.; Shrestha, R.K.; Bai, X.; Singh, M.P.; Lindsey, L.E.; Ren, W. A Global Synthesis of Biochar's Sustainability in Climate-Smart Agriculture—Evidence from Field and Laboratory Experiments. *Renew. Sustain. Energy Rev.* **2023**, *172*, 113042. [CrossRef]
30. Teasley, W.A.; Limmer, M.A.; Seyfferth, A.L. How Rice (*Oryza sativa* L.) Responds to Elevated as under Different Si-Rich Soil Amendments. *Environ. Sci. Technol.* **2017**, *51*, 10335–10343. [CrossRef]
31. *GB/T 42819-2023*; General Technical Code of Practice for Immobilization of Heavy Metal Contaminated Soil in Agriculture Production Area (English Version). Standardization Administration of China: Beijing, China, 2023. Available online: https://www.codeofchina.com/standard/GBT42819-2023.html (accessed on 7 September 2024).
32. Renaud, M.; Chelinho, S.; Alvarenga, P.; Mourinha, C.; Palma, P.; Sousa, J.P.; Natal-da-Luz, T. Organic Wastes as Soil Amendments—Effects Assessment towards Soil Invertebrates. *J. Hazard. Mater.* **2017**, *330*, 149–156. [CrossRef]
33. Tao, C.; Li, R.; Xiong, W.; Shen, Z.; Liu, S.; Wang, B.; Ruan, Y.; Geisen, S.; Shen, Q.; Kowalchuk, G.A. Bio-Organic Fertilizers Stimulate Indigenous Soil Pseudomonas Populations to Enhance Plant Disease Suppression. *Microbiome* **2020**, *8*, 137. [CrossRef]
34. Buss, W.; Hasemer, H.; Ferguson, S.; Borevitz, J. Stabilisation of Soil Organic Matter with Rock Dust Partially Counteracted by Plants. *Glob. Change Biol.* **2024**, *30*, e17052. [CrossRef]
35. Jones, M.E.; LaCroix, R.E.; Zeigler, J.; Ying, S.C.; Nico, P.S.; Keiluweit, M. Enzymes, Manganese, or Iron? Drivers of Oxidative Organic Matter Decomposition in Soils. *Environ. Sci. Technol.* **2020**, *54*, 14114–14123. [CrossRef]
36. Li, X.-M.; Chen, Q.-L.; He, C.; Shi, Q.; Chen, S.-C.; Reid, B.J.; Zhu, Y.-G.; Sun, G.-X. Organic Carbon Amendments Affect the Chemodiversity of Soil Dissolved Organic Matter and Its Associations with Soil Microbial Communities. *Environ. Sci. Technol.* **2019**, *53*, 50–59. [CrossRef]
37. Li, Y.; Wang, T.; Camps-Arbestain, M.; Suárez-Abelenda, M.; Whitby, C.P. Lime and/or Phosphate Application Affects the Stability of Soil Organic Carbon: Evidence from Changes in Quantity and Chemistry of the Soil Water-Extractable Organic Matter. *Environ. Sci. Technol.* **2020**, *54*, 13908–13916. [CrossRef] [PubMed]
38. Ji, Y.; Zheng, N.; An, Q.; Wang, S.; Sun, S.; Li, X.; Chen, C.; Sun, S.; Jiang, Y. Enhanced Immobilization of Cadmium and Lead in Contaminated Soil Using Calcium Alginate-Modified HAP Biochar: Improvements in Soil Health and Microbial Diversity. *Environ. Pollut.* **2024**, *357*, 124445. [CrossRef] [PubMed]
39. Wang, R.; Xu, Z.; Chen, S.; Su, J.; Huang, Y.; Hu, Y. Tradeoffs between pH, Dissolved Organic Carbon, and Mineral Ions Regulate Cadmium Uptake by *Solanum* Hyperaccumulators in Calcareous Soil. *Environ. Res.* **2024**, *248*, 118393. [CrossRef] [PubMed]
40. Liu, C.; Lin, H.; Li, B.; Dong, Y.; Yin, T.; Chen, X. Endophyte Inoculation Redistributed Bioavailable Cd and Nutrient in Soil Aggregates and Enhanced Cd Accumulation in *Phytolacca acinosa*. *J. Hazard. Mater.* **2021**, *416*, 125952. [CrossRef]
41. Zhou, H.; Wang, Z.-Y.; Li, C.; Yuan, H.-W.; Hu, L.; Zeng, P.; Yang, W.-T.; Liao, B.-H.; Gu, J.-F. Straw Removal Reduces Cd Availability and Rice Cd Accumulation in Cd-Contaminated Paddy Soil: Cd Fraction, Soil Microorganism Structure and Porewater DOC and Cd. *J. Hazard. Mater.* **2024**, *476*, 135189. [CrossRef]
42. Wang, X.; Zhai, X.; Lian, J.; Cheng, L.; Wang, M.; Huang, X.; Chen, Y.; Pan, J.; He, Z.; Yang, X. Varietal Responses to a Soil Amendment: Balancing Cadmium Mitigation and Mineral Biofortification in Wheat Production. *Sci. Total Environ.* **2024**, *926*, 171772. [CrossRef]
43. Lian, J.; Cheng, L.; Wang, X.; Chen, Y.; Deng, C.; Wang, Y.; Pan, J.; Shohag, M.J.I.; Xin, X.; He, Z.; et al. Bespoke ZnO NPs Synthesis Platform to Optimize Their Performance for Improving the Grain Yield, Zinc Biofortification, and Cd Mitigation in Wheat. *ACS Sustain. Chem. Eng.* **2024**, *12*, 716–727. [CrossRef]
44. Sarwar, G. Available Amino Acid Score for Evaluating Protein Quality of Foods. *J. Assoc. Off. Anal. Chem.* **1984**, *67*, 623–626. [CrossRef]
45. Huang, X.; Wang, X.; Liu, X.; Cheng, L.; Pan, J.; Yang, X. Nanotechnology in Agriculture: Manganese Ferrite Nanoparticles as a Micronutrient Fertilizer for Wheat. *Plants* **2024**, *13*, 1395. [CrossRef]

plants

MDPI

Article

Application of Biochar-Immobilized *Bacillus megaterium* for Enhancing Phosphorus Uptake and Growth in Rice

Keru Yu [1,2,3,†], Zhenyu Wang [1,3,†], Wenyan Yang [4], Shuai Li [1,3], Dongtao Wu [5], Hongtao Zheng [1,3], Zhengqian Ye [1,3], Shaona Yang [6,*] and Dan Liu [1,3,*]

1 State Key Laboratory of Subtropical Silviculture, Zhejiang A&F University, Hangzhou 311300, China; m17866926886@163.com (Z.W.); ls1851325778@163.com (S.L.); m18376727861@163.com (H.Z.); yezhq@zafu.edu.cn (Z.Y.)
2 School of Environmental and Resources, Zhejiang A&F University, Hangzhou 311300, China
3 Key Laboratory of Soil Contamination Bioremediation of Zhejiang Province, Zhejiang A&F University, Hangzhou 311300, China
4 College of Tea Science and Tea Culture, Zhejiang A&F University, Hangzhou 311300, China; wyyang1808@163.com
5 Soil Fertilizer and Plant Protection and Energy Sources Station of Lishui City, Hangzhou 323000, China; wudongtao1@163.com
6 Zhoushan Agricultural Technology Extension Center, Zhoushan 316021, China
* Correspondence: ysn473@sina.com (S.Y.); liudan@zafu.edu.cn (D.L.)
† These authors contributed equally to this work.

Abstract: Phosphorus (P) is an essential nutrient for rice growth, and the presence of phosphate-solubilizing bacteria (PSB) is an effective means to increase soil P content. However, the direct application of PSB may have minimal significance due to their low survival in soil. Biochar serves as a carrier that enhances microbial survival, and its porous structure and surface characteristics ensure the adsorption of *Bacillus megaterium*. Inoculating rice husk biochar-immobilized with *Bacillus megaterium* (BMB) resulted in dissolved inorganic and organic P levels of 39.55 and 31.97 mL L^{-1}, respectively. Subsequently, rice pot experiments were conducted to investigate the response of soil microbial P mobilization and P uptake in rice to fertilizer inputs. The organic fertilizer (OF) combined with BMB treatment (MOF) showed the highest soil available phosphorus (AP) at 38 days, with a value of 7.83 mg kg^{-1}, as well as increased the *pqqC* abundance while decreasing the abundance of *phoD* bacterial communities compared with the control. Furthermore, the bioavailable P reservoir (H_2O–Pi and $NaHCO_3$–Pi) in soil was greatly increased through the fertilizer input and microbial turnover, with the highest H_2O–Pi (3.66 mg kg^{-1}) in OF treatment and the highest $NaHCO_3$–Pi (52.65 mg kg^{-1}) in MOF treatment. Additionally, carbon utilization analysis was applied using the commercial Biolog system, revealing that the MOF treatment significantly increased the utilization of carbohydrates, polymers, and amino acid carbon sources. Moreover, compared to the control, MOF treatment significantly increased the shoot (0.469%) and root P (0.516%) content while promoting root development and thereby supporting rice growth. Our study demonstrates that the MOF treatment displayed higher P levels in both soil and rice plants, providing a theoretical basis for further understanding the role of biochar-based bacterial agents in rice P management.

Keywords: rice husk biochar; phosphate solubilizing bacteria; phosphorus gene community; microbial catabolic activity; phosphorus content

1. Introduction

Rice is a major food source, particularly in Asia [1], with adequate phosphorus (P) being essential for its root development and maturation [2]. However, soil pH, mineral composition and other physiochemical factors can limit phosphorous availability [3]. As a result, rice cultivation practices often rely on the application of large amounts of phosphorus fertilizer to ensure that it does not become a limiting factor in rice production. However, PO_4^{3-} is easily fixed by soil particles, making it difficult for crops to absorb and utilize, leading to low phosphorus utilization efficiency [4]. Furthermore, long-term application of phosphorus fertilizer not only results in the wastage of this non-renewable mineral resource [5] but also in soil phosphorus pollution, soil degradation, compaction and eutrophication of water bodies [6]. Therefore, it is crucial to activate immobilized phosphorus in paddy soil to enhance the efficiency of rice phosphorus utilization.

Phosphate-solubilizing bacteria (PSB) are widely used in agriculture as a biofertilizer to improve phosphorus absorption and crop yields [7]. These bacteria have the ability to secrete organic acids, which can chelate Ca, Fe, and Al, leading to the release of phosphorus [2]. Specifically, PSB expresses pyrroloquinoline quinone (PQQ), a key enzyme in gluconic acid synthesis that promotes the release of inorganic phosphorus (Pi) [8]. The *pqqC* gene is commonly used as a bioindicator for the presence of Pi-solubilizing microorganisms and plays an important role in this process [9]. Moreover, microorganisms are essential in facilitating the accumulation of phosphorus in plants by mineralizing organic phosphorus (Po) via enzyme action [10]. Detection of *phoD* gene encoding alkaline phosphatase is also used to assess changes in bacterial communities driving Po mineralization [8,11]. Inoculating PSB is considered to be an effective way to improve soil available phosphorus (AP) content and achieves high phosphorus nutrient efficiency. However, practical application is hindered by the low survival rate of these bacteria, limiting the effectiveness of PSB's benefits [12]. Therefore, finding a highly stable and robust carrier material could enhance the overall availability of phosphorus nutrients.

Biochar, a carbon-rich material produced by the pyrolysis of biomass under anaerobic conditions [13], is applied as a soil amendment to enhance phosphorus cycling in soil–plant systems [14]. It is widely recognized as an excellent carrier material due to its unique physical and chemical characteristics that provides a protective habitat for inoculated microbial communities and have shown positive effects on increasing microbial abundance in soils [15,16]. Previous studies have demonstrated that the porous structure of biochar enhances PSB adsorption capacity [17]. Furthermore, biochar contains a large amount of nutrients such as organic carbon, phosphorus and nitrogen, along with numerous high-energy adsorption sites that provides substrates and reaction sites for bacteria immobilization [18,19].

Based on our previous soil incubation experiment, we found that the combination of 1% (*w*/*w*) biochar-immobilized *Bacillus megaterium* (BMB) and 0.1% OF resulted in higher soil AP compared to the combination of 0.5% BMB and 0.1% OF. In this study, rice plants were cultivated in the presence of either 1% Rice husk biochar (RHB) or BMB and 0.1% OF, which were applied to the soil in order to investigate: (1) responses of different phosphorus fractions to different fertilization regimes (2) responses of bacterial communities harboring *pqqC* and *phoD*, as well as their associated carbon metabolism (3) relationships among soil properties, phosphorus fractions, *pqqC*- and *phoD*-harboring bacterial communities, and metabolic activity of carbon substrate and (4) factors influencing phosphorus uptake in rice. We hypothesized that different fertilization regimes would alter soil phosphorus fractions and accelerate its conversion via the abundance of the inoculated microorganisms. An increase in AP content and uptake results in increased rice growth. The findings of this study could contribute to a better understanding of the mechanisms involved in soil P cycling, as well as provide a theoretical basis for improving the utilization rate of P in paddy soil and promoting rice growth.

2. Results

2.1. Characterization Analyses of RHB and BMB

SEM images revealed that the RHB surfaces were predominantly microporous, with *Bacillus megaterium* appearing as an elliptical rod-like structure distributed on the biochar surface. Interestingly, no obvious surface structure alterations were observed on the biochar due to its association with bacteria (Figure S1a,c). EDS images indicated that RHB primarily consisted of biomass material composed of C and O, along with associated N, K, Ca, S and P elements. Moreover, loading of the bacteria (BMB) resulted in decreases of each element within the biochar material (Figure S1b,d).

Our RHB and BMB samples were analyzed using FTIR, indicating a high degree of similarity in terms of functional groups (Figure S2). Both exhibited characteristic peaks at 3423 and 3425 cm^{-1} (–OH stretching vibrations) and 2815 and 2814 cm^{-1} (asymmetric stretching vibrations of $-CH_2$ in fatty acids), respectively. The presence of peaks at 1593 cm^{-1} indicated the existence of Carboxyl (C=O) groups, while multiple absorption peaks at 680–1450 cm^{-1} were also observed (Figure S2). These data suggested that bacterial inoculation did not alter the chemical composition or structure of the biochar.

2.2. Changes in BMB Phosphorus Solubilizing Capacity

We then investigated whether BMB samples exhibited significant ability to mobilize phosphorus (Figure S3). In the inorganic phosphorus medium, the AP content increased rapidly during the first 1–2 d after inoculation and reached its peak on day 3 (29.55 mg L^{-1}), remaining relatively stable at later stages of culture. Additionally, BMB demonstrated considerable solubilizing capacity for organic phosphorus. The AP content rose rapidly from days 1 to 3, reaching its maximum on day 4 (31.97 mg L^{-1}), and then stabilized.

2.3. Changes in Soil Properties

We initially investigated changes in soil properties over time (Table S2). The data indicated that within the first 24 h (day 0), the soil pH, AP and TP increased in all treatments compared to the control. The results showed that the experimental treatment had a significant effect on soil pH value and phosphorus content in the short term, which may be caused by additives. At day 21, the MOF combination treatment led to the greatest increase in soil pH (6.63) and AP content (14.46 mg kg^{-1}) and was significantly higher than the control. Additionally, soil acid phosphatase activity in the OF treatment significantly increased compared to other treatments ($p < 0.05$). At day 38, all treatments enhanced soil pH, AP content and acid phosphatase activity compared to the control. Notably, the MOF treatment exhibited significant increases in pH (6.19) and acid phosphatase activity (0.282 mg $g^{-1}\cdot 24$ h) ($p < 0.05$).

2.4. Changes in Soil Phosphorus Fractionation

The presence of phosphorus in the soil after 38 d indicated that all treatments significantly increased levels of H_2O–P ($p < 0.05$) compared to the control, with increases of 35.03–59.60% (Figure 1a). The MOF treatment showed significantly higher levels of $NaHCO_3$–Pi (52.65 mg kg^{-1}) and NaOH–Pi (168.15 mg kg^{-1}) compared to the control ($p < 0.05$), with increases of 8.71% and 11.61%, respectively (Figure 1b,d). There was no significant difference in $NaHCO_3$–Po, NaOH–Po and HCl–Pi content among the different treatments ($p > 0.05$) (Figure 1c,e,f).

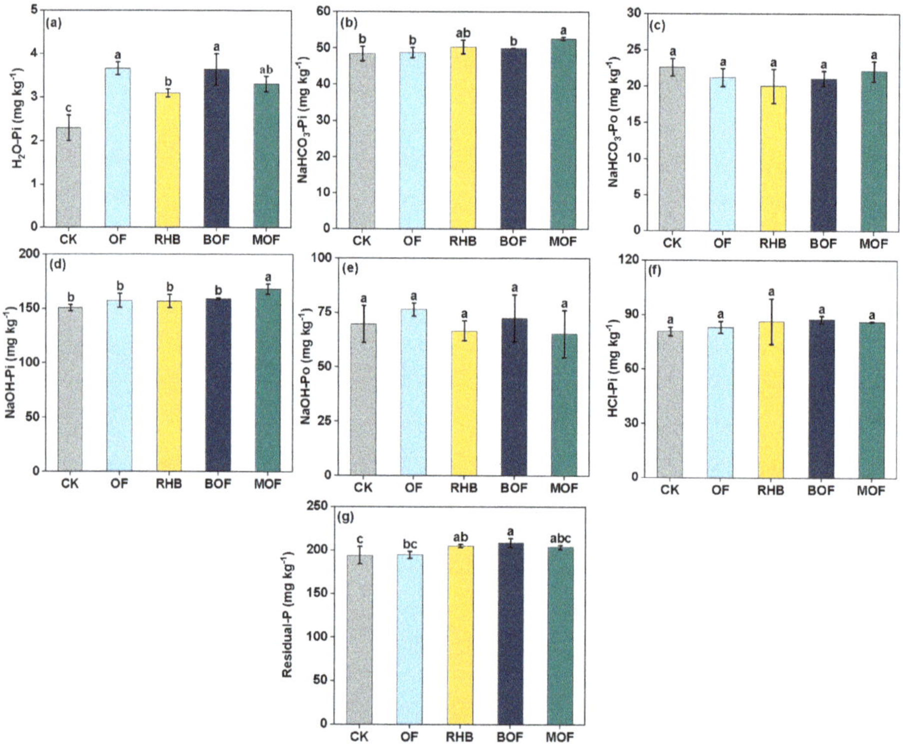

Figure 1. Content of different phosphorus fractions in soil on day 38. (**a**) H_2O–P (**b**) $NaHCO_3$–Pi (**c**) $NaHCO_3$–Po (**d**) NaOH–Pi (**e**) NaOH–Po (**f**) HCl–P (**g**) Residual–P. Error bars represent the standard deviation of the mean (n = 3). Different letters within a column indicate significant differences at $p < 0.05$. CK, control without fertilizer; OF; 0.1% organic fertilizer; RHB, 1% rice husk biochar; BOF, 1% rice husk biochar and 0.1% organic fertilizer. MOF, 1% rice husk biochar-immobilized Bacillus megaterium (BMB) and 0.1% organic fertilizer.

2.5. Response of the Soil pqqC- and phoD-Harboring Bacterial Communities

The abundance and diversity of *phoD*- and *pqqC*-harboring communities were influenced by the different fertilization regimes (Figure S4). Compared to the control, the Shannon diversity index of the *pqqC* community was significantly changed in the OF treatment, which increased by 22.38% (Figure S4a). Additionally, the Chao1 abundance index of the *pqqC* community was significantly affected in both the BOF and MOF treatments, increasing by 55.69% and 50.25%, respectively ($p < 0.05$) (Figure S4b). Compared with the control, all treatments decreased the Shannon diversity and Chao1 abundance indices of *phoD* communities, with reductions ranging from 3.38–12.29% and 18.98–36.14%, respectively, while BOF treatment decreased significantly (Figure S4a,b).

Under all fertilization treatments, BOF and RHB treatments showed the maximum relative abundance of the dominant phylum *Proteobacteria* in the bacterial community carrying *phoD* and *pqqC* genes, which was 77.20% and 91.37%, respectively. (Figure 2a,b). In addition, *Actinobacteria* accounted for a considerable proportion of the *pqqC* gene-carrying bacterial communities. Especially under the treatment OF, the relative abundance of *Actinobacteria* reached the highest value, 32.31% (Figure 2a). The predominant genera across all samples were *Noviherbaspirillum* (4.75–11.75%), *Pelomonas* (1.73–17.15%) and *Mycobacterium*

(5.12–12.08%) for the *pqqC* community (Figure 2c). RHB showed the highest abundance of *Noviherbaspirillum* and *Pelomonas*, while the MOF treatment showed the lowest abundance of *Mycobacterium* (Figure 2c). Moreover, *Bradyrhizobium* (26.73–41.40%), *Ramlibacter* (4.53–15.58%), and *Microcoleus* (6.52–13.76%) predominated in the *phoD* communities. The BOF treatment exhibited the highest abundance of *Bradyrhizobium* (41.40%), whereas the OF treatment possessed the highest abundance of *Ramlibacter* (15.58%) compared to other treatments (Figure 2d).

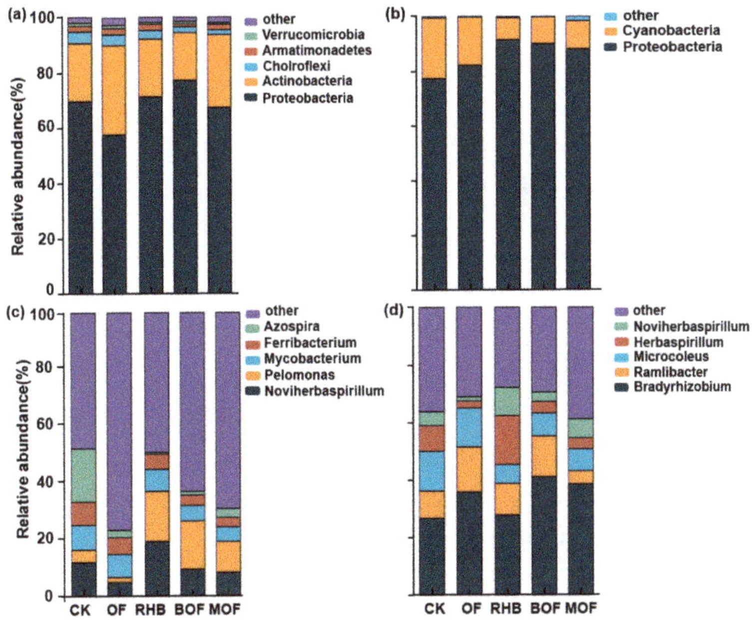

Figure 2. Relative abundance at the phylum and genus levels in (**a**,**c**) *pqqC* and (**b**,**d**) *phoD* community composition for different treatments. Abundance > 0.5% for phyla and > 1.0% for genus, respectively. CK, control without fertilizer; OF; 0.1% organic fertilizer; RHB, 1% rice husk biochar; BOF, 1% rice husk biochar and 0.1% organic fertilizer. MOF, 1% rice husk BMB and 0.1% organic fertilizer.

2.6. Response of the Soil Microbial Activity

The effects of different treatments on soil microbial activity were assessed based on the level of color development (AWCD) in the Biolog plates. Changes in AWCD over the 168 h incubation period revealed distinct differences between MOF and the other treatments, with these variances becoming more pronounced over time (Figure 3a). Following a 96 h incubation, MOF treatment showed a significant increase in total AWCD for all 31 carbon substrates ($p < 0.05$), with a 4.70-fold rise in total AWCD when compared with the control (Figure 3b). Specifically, the MOF treatment samples significantly increased the utilization of carbohydrates, polymers, and amino acids, while BOF and MOF treatments showed significant increases in the utilization of carboxylic acids. And there were no significant differences in the utilization of Phenols or Amines ($p > 0.05$) (Figure S5).

The utilization of the 31 carbon substrates after 96 h incubation differed among our experimental treatments, with the biggest differences observed between the MOF and control (Figure 4). The utilization of carboxylic acids and amino acids in MOF treatment was much higher than that in other treatments. In addition, the utilization efficiency of carbohydrates (β-methyl-D-glucoside, D-mannitol, D-cellulobiose and D-galactolactone) also reached the highest in MOF treatment. Moreover, i-erythritol, N-acetyl-D-glucosammine, α-cycogen, and 2-hydroxybenzoic acid were higher in the CK treatment compared to other treatments. Furthermore, *D-xylose* in RHB treatment and D-galactonic acid lactone in BOF treatment were at higher levels when compared to other treatments.

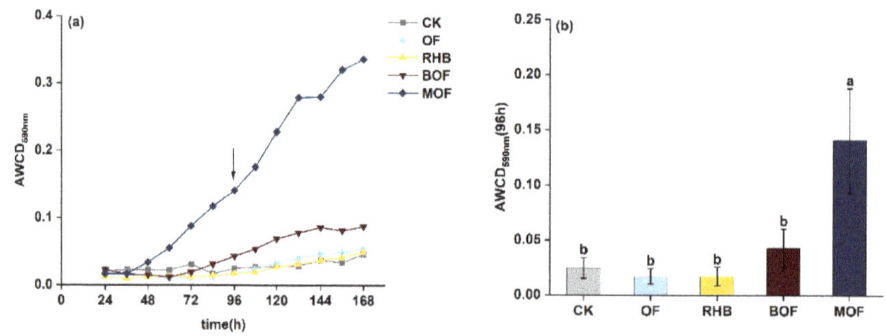

Figure 3. Dynamics of carbon source utilization in EcoPlates. (**a**) Average well color development (AWCD) index of the soil samples and (**b**) Total AWCD after 96 h. Error bars represent the standard deviation of the mean (n = 3). Different letters within a column indicate significant differences at $p < 0.05$. The black arrow indicates the sampling time (96 h) chosen for further analysis. CK, control without fertilizer; OF; 0.1% organic fertilizer; RHB, 1% rice husk biochar; BOF, 1% rice husk biochar and 0.1% organic fertilizer. MOF, 1% rice husk BMB and 0.1% organic fertilizer.

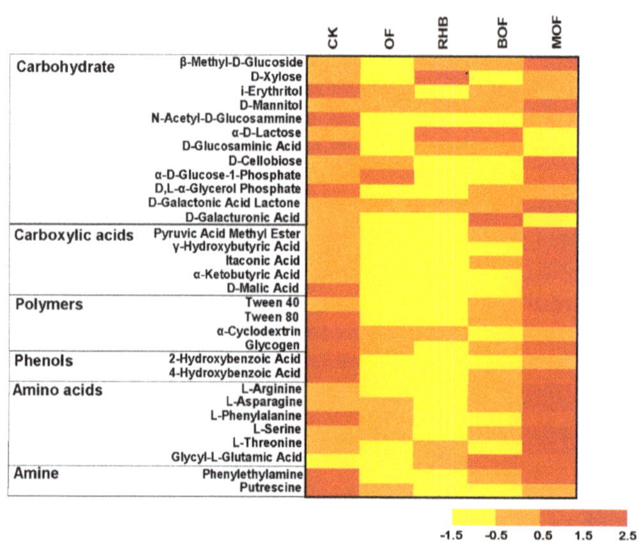

Figure 4. Biolog EcoPlate substrate utilization (Abs 590 nm) after 96 h of incubation. CK, control without fertilizer. OF, 0.1% organic fertilizer. RHB, 1% rice husk biochar. BOF, 1% rice husk biochar and 0.1% organic fertilizer. MOF, 1% rice husk BMB and 0.1% organic fertilizer.

2.7. Plant Evaluation of Rice Growth Promotion

Phosphorus accumulation and plants' dry weight were influenced differently by the various fertilization regimes. When compared with the control, SPAD values for all other treatments increased in two stages, and the MOF samples showed significant increases of 11.69% and 13.07%, respectively ($p < 0.05$) (Figure 5a). MOF treatment generated the highest values (0.469%) with significant increases of 8.31% on day 38 compared to the control (Figure 5d). In addition, the fertilization treatments had a positive effect on rice root growth. It can be seen that at 21 and 38 d, rice root dry weight and phosphorus content for MOF treatment were significantly ($p < 0.05$) higher than that of control and showed increases of 36.77 and 9.52% (1.21- and 0.43-fold), respectively (Figure 5c,e).

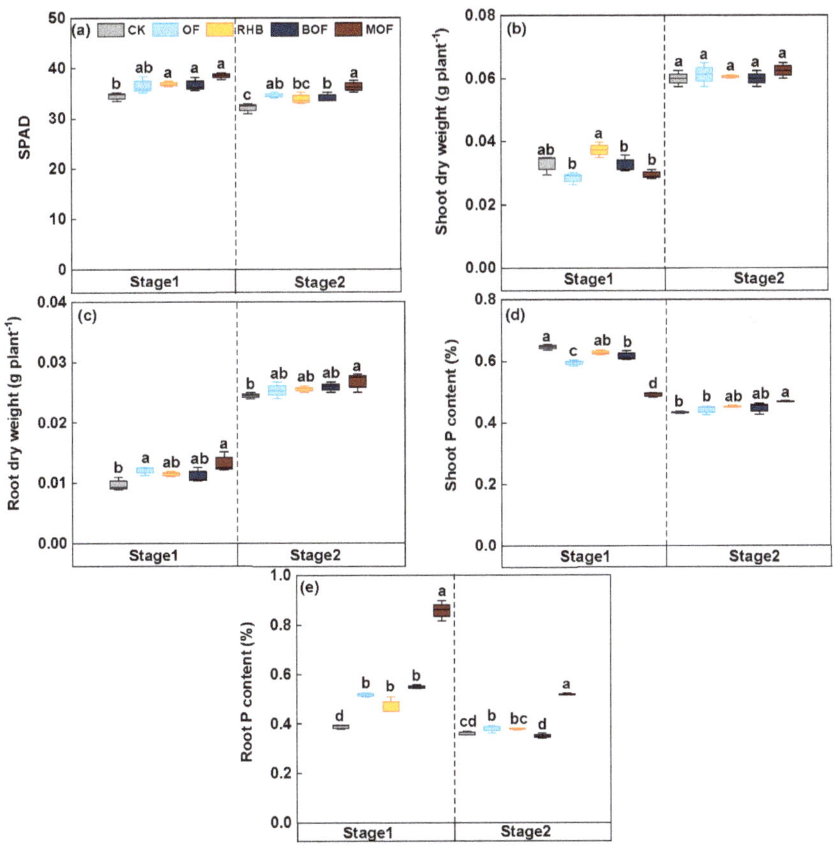

Figure 5. Rice growth parameters. (**a**) SPAD values (**b**) Shoot dry weight (**c**) Root dry weight (**d**) Shoot P content (**e**) Root P content. Stage 1 and Stage 2 represent days 21 and 38, respectively. Means ± standard deviations for three replicates. Different letters within a column indicate significant differences at $p < 0.05$. CK, control without fertilizer; OF; 0.1% organic fertilizer; RHB, 1% rice husk biochar; BOF, 1% rice husk biochar and 0.1% organic fertilizer. MOF, 1% rice husk BMB, and 0.1% organic fertilizer.

Compared to the control group, all treatments generated positive effects on root indices (Table S3). The total root length was significantly increased by 11.30–45.24% for the MOF treatment compared to the other treatments ($p < 0.05$). Furthermore, compared to the control group, the MOF treatment significantly increased root volume (23.2%), root projection area (33.16%), root surface area (33.20%) and the number of forks (43.68%) ($p < 0.05$). The above results show that the positive effect of MOF on root growth was demonstrated.

2.8. Relationships Among Soil Properties, Microbial Activity and Communities of pqqC and phoD

Principal component analysis (PCA) showed that PCA1 and PCA2 explained 27.97% and 19.79% of the variation in *pqqC* and 36.71% and 15.08% of the variation in *phoD*, respectively (Figure 6a,b). Control samples were mainly distributed in PCA1-positive regions, while different treatments caused the distribution of *pqqC* gene-carrying bacterial communities to shift to PCA1-negative regions (Figure 6a). Similarly, different treatments also caused the distribution of *phoD* gene-carrying bacterial communities to shift to PCA1-negative regions (Figure 6b). Figure 6c analysis was applied to reveal discriminations between various experimental treatments. The contributions of PCA1 and PCA2 accounted for 35.33 and 19.24%, respectively, to the variance of carbon source substrate. The control samples were primarily located in the negative zone of PCA1, while the different treatments caused a shift toward the positive zone. This analysis also revealed strong interrelations between soil properties such as pH, AP, and acid phosphatase, as well as carbohydrates, carboxylic acids, and polymers after 96 h of incubation.

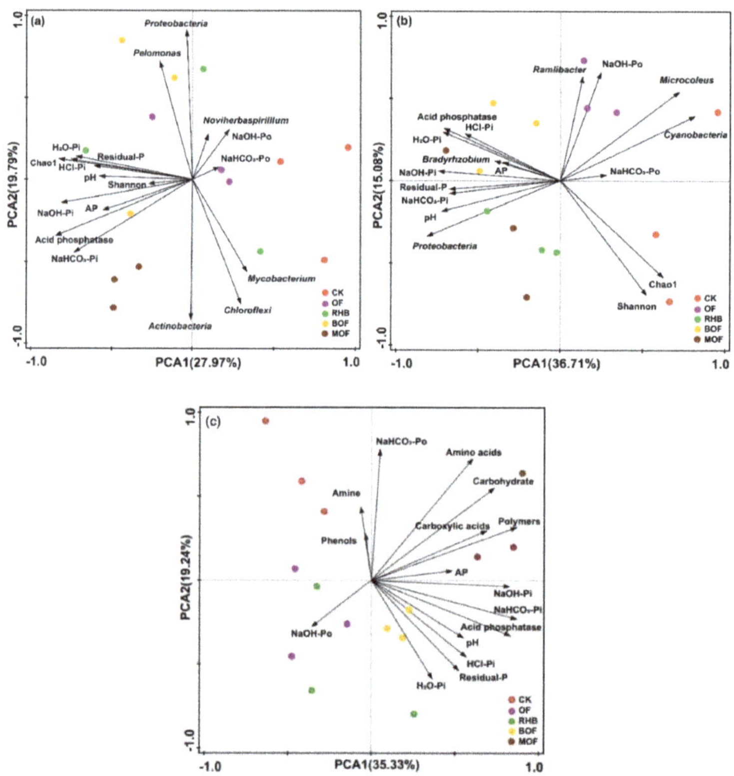

Figure 6. Principal component analysis (PCA) plots showing the relationships between soil properties and the diversity of the (**a**) *pqqC* and (**b**) *phoD* bacterial communities and (**c**) six groups of substrates (carbohydrates, carboxylic acids, polymers, phenols, amino acids, amines) after 96 h of incubation. Control, OF, RHB, BOF and MOF are indicated by red, purple, green, yellow and brown circles, respectively. The percentage of contribution is shown for each axis. CK, control without fertilizer; OF; 0.1% organic fertilizer; RHB, 1% rice husk biochar; BOF, 1% rice husk biochar and 0.1% organic fertilizer. MOF, 1% rice husk BMB and 0.1% organic fertilizer.

2.9. Influential Factors on Rice Phosphorus Uptake

We conducted the Mantel test to identify the factors affecting phosphorus uptake in rice plants (Figure 7). The shoot phosphorus content showed a significantly positive correlation with AP, NaOH–Pi, NaOH–Po, AWCD, carbohydrates, polymers, and amino acids ($p < 0.05$). Moreover, the root phosphorus content exhibited a highly significantly positive correlation with $NaHCO_3$–Pi and NaOH–Pi, AWCD, carbohydrates, polymers, and amino acids ($p < 0.01$).

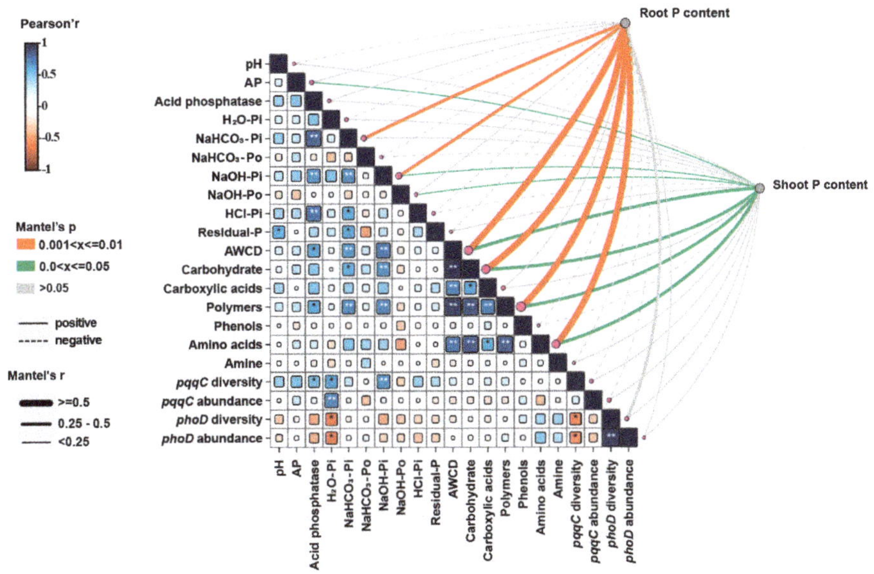

Figure 7. Pearson's correlation matrix between variables under pot experiment(* $p < 0.05$, ** $p < 0.01$). The color density gradient and shape size represents the correlation coefficients, as indicated. The Mantel test was used to analyze the relationship between soil and plant variables and P contents in rice tissues. Line width corresponds to the Mantel R statistic, and line color relates to correlation.

3. Discussion

3.1. Physico-Chemical Properties of BMB and Phosphorus Solubilization

Biochar is widely recognized as an excellent carrier material due to its porous structure [15]. It supports microbial habitats and contributes to facilitating PSB attachment and proliferation [20,21]. The SEM–EDS images of RHB and BMB presented here demonstrated its unique pore structure and chemical elements, which enabled the distribution of BM on the surface and within pores without altering the properties of the supporting biochar (Figure S1a,b). Meanwhile, the presence of Hydroxy, Carboxyl and Amine groups on biochar surfaces promotes bacterial adhesion as well as proliferation [17]. Thus, the dominant substances of biochar are available for PSB and affect the load of PSB.

BMB exhibited a high solubilization ability in growth media containing both organic and inorganic phosphorus, which had been attributed to phosphorus-solubilizing bacteria (BM) by secreting organic acids to reduce the pH of the media, thereby increasing the solubility of phosphorus (Figure S3) [22]. Additionally, biochar functions as a neutralizing agent, enhancing the acid production capacity of PSB and improving the adaptability of PSB to adverse environments [17,23]. It is worth noting that biochar contains functional groups such as Hydroxyls and Carboxyls, which actively contribute to phosphorus solubilization [24]. Therefore, the existence of BM promoted the P-solubilizing effect.

3.2. Soil Properties and Phosphorus Fractions

Low phosphorus availability is a key limiting factor for rice growth [25]. Numerous studies have demonstrated that biochar coupled with organic fertilizer significantly improved soil properties, including pH, nutrient storage and enzyme activities [18,26,27]. This is consistent with our research results, on the 21st day of the experiment, the pH and AP contents of MOF treatment were significantly increased compared with control, increasing by 8.16% and 44.74%, respectively. In our study, the application of biochar increased soil pH primarily due to the dissolution of alkaline substances (such as inorganic carbonate) in the biochar [28]. At the same time, the dynamic balance of active phosphorus, that is, the permanent exchange of dissolved phosphorus through the adsorption process, and the adsorbed phosphorus entering and leaving the soil solution according to factors such as pH value, temperature, humidity and concentration, determines the balance of phosphorus in the soil pores and the soil's phosphorus buffer capacity [29,30]. Soil phosphorus availability tended to increase when the pH transitions from acidic conditions to a range of pH 6–7 (Table S2). For example, at the beginning of the experiment (day 0), soil pH in the RHB, BOF, and MOF treatment groups was significantly increased by 0.4 to 0.69 units compared to the control group, while available phosphorus (AP) content was also significantly increased by 18.59% to 49.74%. This is consistent with the results of the previous study [31]. The highest P content in MOF treatment indicated that PSB plays a role in increasing soluble phosphorus levels by producing organic acids, protons and hydroxide ions to dissolve minerals [6]. P may be readily soluble (e.g., a soluble compound like a phosphate fertilizer) or found in large organic molecules, which are subject to the rates of microbial organic matter decomposition [32]. Phosphatase activity also increased in all treatments compared with control, indicating high energy demand for microbes mobilizing organic phosphorus compounds [33]. Therefore, based on our findings, we speculate that the application of biochar coupled with organic fertilizer might stimulate the growth of PSB and subsequently promote AP release.

Phosphorus fractional analysis provides an additional method for investigating phosphorus transport and transformation in soils. Our results indicated that levels of H_2O–Pi and $NaHCO_3$–Pi for all treatments were increased, particularly for MOF (Figure 1a,b). Biochar application mainly impacts the soil's inorganic phosphorus component, leading to increase phosphorus availability through concentration gradient diffusion [34]. Furthermore, the application of organic fertilizer might also increase soil organic carbon content, thereby reducing phosphorus adsorption and increasing desorption [8]. It is an important factor of P conversion in soil [35]. We observed a significant positive correlation between phosphatase activity and both $NaHCO_3$–Pi and NaOH–Pi (Figure 7), which is consistent with the findings of Wang et al. [36]. The decrease in $NaHCO_3$–Po and NaOH–Po content may be attributed to the hydrolysis of soil Po into Pi via extracellular phosphatases produced by BM [37]. Compared to the BOF treatment, a reduction in Residual–P content with the MOF treatment suggested that BM could promote the transition from stable to unstable phosphorus (Figure 4). This is mainly the conversion of insoluble P components into bioavailable phosphate by phosphate-solubilizing bacteria [38].

3.3. *pqqC*- and *phoD*-Harboring Bacterial Community

Biochar acts as a slow-release fertilizer and stores nutrients that promotes microbial growth [39]. In addition, it is possible that the inoculant could not survive better because of competition with native microorganisms [40], so the nutrients stored in biochar have a positive effect on increasing the community. Our results revealed that biochar application led to improvements in the diversity and abundance of *pqqC*-harboring bacteria compared with the control. BOF and MOF treatments significantly affected the Chao1 abundance index of the *pqqC* community, increasing by 55.69% and 50.25%, respectively ($p < 0.05$) (Figure S4b). The available organic carbon provided by biochar and organic fertilizers enters the soil environment, supplying external carbon for bacteria. This stimulation contributed to their growth and reproduction, ultimately enhancing bacterial diversity [8,41]. Notably, the diversity and abundance of *pqqC* in MOF treatment were lower than that in BOF treatment (Figure S4b). This difference is likely due to increased BM survival with biochar supports as it became dominant [12]. In contrast, the *phoD* diversity and abundance across various fertilization treatments were decreased. The Shannon diversity and Chao1 abundance indexes decreased by 3.38–12.29%, and 18.98–36.14%, respectively (Figure S4a,b), and this was similar to the findings of previous research [42]. It is possible that in low-nutrient soils, the absorption potential of *phoD*-harboring bacteria was inhibited due to competition for carbon sources and nutrients, leading to decreased abundance.

Consistent with previous studies, *Proteobacteria* and *Actinobacteria* were the dominant phyla in the *pqqC* and *phoD* populations [26,43]. At the same time, the result also indicated the importance of *Proteobacteria* and *Actinobacteria* in P-mobilizing microorganisms [8,44]. *Proteobacteria* and *Actinobacteria* are known as copiotrophic bacteria, and the application of biochar and organic fertilizer provided rich nutrients for stimulating growth and increasing their relative abundance, consistent with previous findings [26,41]. We found that *Bradyrhizobium* had a higher relative abundance, whereas soil AP content was higher. *Bradyrhizobium* is identified as the keystone flore in regulating the compositions and functions of the *phoD*-harboring communities [26] and produces phosphatase or organic acid to increase AP content [45]. Meanwhile, the presence of biochar and organic fertilizer might provide sufficient organic carbon supply for *Bradyrhizobium* survival. Among them, the abundance of *Bradyrhizobium* treated with BOF and MOF was as high as 41.40% and 38.54% (Figure 2d). Thus, OF and BMB applications promoted the formation of highly efficient P mobilization communities and enhanced their potential for Po mineralization and Pi solubilization.

3.4. Microbial Activity

In the present study, significant differences were observed in carbon source metabolism among different treatments. The microbial groups in the MOF treatment exhibited the highest capacity for carbon source utilization, as reflected by their high AWCD scores. After 96 h of incubation, MOF treatment showed a 4.70-fold increase in total AWCD compared to the control group ($p < 0.05$) (Figure 3a,b). Organic carbon serves as the primary source of microbial growth [8]. Meanwhile, the metabolism and reproduction of the soil microbial community can be promoted by increasing soil nutrients [46]. The addition of biochar and organic fertilizer in our study provided fresh exogenous organic carbon and nutrient elements for BM and phosphorus-mobilizing microorganisms were significantly positively correlated with multiple carbon sources (Figure 7).

The utilization of carbohydrates, carboxylic acids, polymers, and amino acid carbons was significantly increased by 3.12, 1.41, 5.95 and 0.86 units in the MOF treatment compared to the control (Figure S5). Although Carboxylic acid carbon sources constitute a minor fraction of soil organic matter, they serve as important C and N sources for microbial respiration and utilization, thereby contributing to soil nutrient accumulation [47,48]. Soil microorganisms obtain N from the soil organic matter to compensate for the high C: N ratio corrected by biochar, and this effect may explain the increased utilization of Amino acid substrates [49]. In addition, the addition of organic fertilizers or biochar both tended to reduce the utilization of phenolic acids (Figure 4). The increase in soil pH might contribute to the neutralization of Phenolic acids, thereby reducing the microbial utilization of phenolic acids [50]. PCA analysis indicated that the metabolic capacity of soil microorganisms in our control treatment was separate from that of other treatments (Figure 6c). This suggested that fertilization significantly altered the soil environment and consequently affected the characteristics of soil microorganism utilization of exogenous carbon sources. Thus, in the case of high utilization of carbon sources, bacterial metabolic activity was promoted to enhance the phosphorus solubilization ability of BM.

3.5. Effect on Rice Growth

The SPAD value is a well-established indicator for the measurement of leaf chlorophyll [51]. The highest SPAD content in MOF treatment proved that it promoted the growth and development of rice and significantly increased by 11.69% and 13.07% compared with the control in two stages (Figure 5). PSB may stimulate plant growth by regulating signaling pathways involved in photosynthesis [52]. Previous studies have widely reported the beneficial effects of biochar and organic fertilizer on soil nutrient enhancement and plant nutrient uptake [12,53]. Similarly, our results indicated that MOF treatment had a superior effect on plant phosphorus accumulation and phosphorus content was significantly increased by 9.52% compared with the control, suggesting BM inoculation produced the best effect on phosphorus mobilization (Figure 5). In addition, we observed an increase in both biomass and phosphorus content of rice roots relative to shoots. PSB inoculation was accompanied by microbe-induced changes in rice roots [2]. At the same time, biochar could increase soil porosity due to its unique pore structure [54], potentially allowing for greater root penetration. Compared with the control, the positive effect of MOF treatment on root parameters, significantly increasing root volume (23.2%), root protruding area (33.16%), root surface area (33.20%) and root branching number (43.68%) (Table S3). Increased uptake of phosphorus may promote root growth and thus increase nutrient uptake [55]. The maximum root diameter observed under MOF treatment (0.68 mm) aligns with the concept that thicker roots tend to anchor plants while enhancing nutrient transport capacity and promoting plant growth better [56].

4. Materials and Methods

4.1. Experimental Materials

The soil in this experiment was collected from the 0–20 cm depth of the surface layer of Xintong, Fuyang District, Hangzhou City, Zhejiang Province, Eastern China (119°83′ E, 29°91′ N). RHB was provided by Zhejiang Yangtze River Delta Junong Technology. The organic fertilizer (OF) utilized was a straw-mixed fertilizer made from rice straw. *Bacillus megaterium* was obtained from the BeNa Culture Collection and then cultured in a nutrient broth (NB) medium. The solid immobilized bacterial agent—BMB was obtained according to the method of Zhu et al. [57]. In brief, bacterial seed solution was produced by shaking at 160 rpm until logarithmic growth. Subsequently, 5 g of biochar that had been passed through a 100 mesh sieve was added into 100 mL NB and sterilized at 121 °C for 20 min. Then, the NB with inoculated bacterial seed solution (2%) was incubated in a shaker set at 160 rpm, 30 °C for 24 h. At last, the culture solution was subjected to centrifugation and washed with a saline solution containing a mass fraction of 1%. The physical and chemical properties of the soil, biochar, and organic matter utilized in this study are presented in Table S1.

4.2. Characterization and the Phosphate-Solubilizing Ability of BMB

Fourier transform infrared (FTIR, Thermo Fisher, Waltham, MA, USA) was performed on samples prepared by the KBr pressed-disk technique. The morphology and structure of RHB and BMB were examined using an SU8010 scanning electron microscope (SEM, Hitachi, Japan), and elemental content was determined by energy dispersive spectrometry (EDS).

Soluble phosphorus was quantified using 0.5 g of BMB inoculated into 100 mL Pi and Po media, with insoluble $Ca_3(PO_4)_2$ and lecithin as the sole phosphorus source. The mixtures were then incubated at 30 °C with shaking at 160 rpm. Flasks lacking added materials or RHB were used as blank controls. After a 7-day incubation period, 2 mL samples were taken every 24 h, and the supernatant was extracted by centrifuging at 12,000 rpm for 10 min, repeated three times. Soluble phosphorus levels were determined using molybdenum blue colorimetry [18].

4.3. Pot Experiments and Sampling

Experimental soil samples were air-dried, ground through a 5 mm sieve, and added to plastic pots (85 × 85 × 108 mm) filled with 500 g of well-mixed air-dried soil. The fertilization regimes, including three replicates of five treatments, were arranged in a completely randomized design. The treatments consisted of: (1) control without fertilizer (CK), (2) organic fertilizer (OF), (3) rice husk biochar (RHB), (4) rice husk biochar and organic fertilizer combined (BOF), as well as (5) rice husk BMB and organic fertilizer combined (MOF). OF, RHB, and rice husk BMB were added at the ratio of 0.1%, 1%, and 1% of soil weight, respectively. The rice seeds (cultivar Shenchengyou 913) were germinated in tap water. After growing for 7 days, they were gently collected and then 10 healthy seeds with similar height were selected for transplantation. The seeds were planted in equal spaces apart from each other in the pot with a maintained water layer of about 2 cm. The experiment was started on 24th October 2023, and top dressing N (20 mg kg^{-1} soil) and K (50 mg kg^{-1} soil) were applied as $(NH_4)_2SO_4$ and KCl, respectively. It was incubated under outdoor temperature conditions and flooded throughout the experiment to simulate the situation of flooded rice. Destructive sampling was conducted at 0, 21, and 38 d after the start of the experiment. A portion of each fresh soil sample was stored at −80 °C, while the remainder was air-dried and passed through 10- and 100-mesh sieves. The plants were then rinsed with deionized water and divided into above-ground (shoot) and below-ground (root) portions. Subsequently, they were dried at 70 °C until constant weight and finely ground for analysis.

4.4. Sample Testing and Soil Phosphorus Fractionation

Soil pH was measured using a pH meter with a soil-to-water ratio of 1:2.5 (m/V). Total phosphorus (TP) was determined by digestion with perchloric acid ($HClO_4$), and AP was determined following 0.5 M $NaHCO_3$ extraction as previously described [58]. Soil acid phosphatase activity was determined by phenylene disodium phosphate colorimetry [59]. Soil phosphorus fraction content was determined using the Hedley sequential extraction procedure modified as previously reported [60]. Briefly, 0.5 g soil was extracted in sequence using 30 mL (0.5 M $NaHCO_3$ pH 8.5, 0.1 M NaOH and 1.0 M HCl) with oscillation for 16 h at 25 °C at 180 rpm. Supernatants of centrifugation were digested with acidified ammonium persulfate. Po was calculated as the difference between total P (Pt) and Pi. Finally, the soil residue was digested with H_2SO_4–H_2O_2 to extract residual phosphorus. Phosphorus levels in all extracts and digestive fluids were determined. Fresh plant roots were analyzed using an Epson V 700 scanner with WinRHIZO Pro software (Reagent

Instruments, Québec City, QC, Canada). Chlorophyll levels in shoots were measured using a Soil and Plant Analyzer Development (SPAD) 502 Plus meter. The dried plant samples were digested with H_2SO_4–H_2O_2 and phosphorus content in the solution was determined with the molybdenum ascorbate blue method [58].

4.5. High-Throughput Sequencing and Data Processing

Genomic DNA was extracted from freeze-dried soil using the MagaBi Soil/Stool DNA Kit (Bioer, Hangzhou, China) according to the manufacturer's instructions. The purity of DNA was determined using a Qubit 4.0 instrument (Thermo Fisher Scientific, Waltham, MA, USA). The amplification primers for bacterial *phoD* and *pqqC* genes were as follows: (5′ to 3′) CAGTGGGACGACCACGAGGT/GAGGCCGATCGGCATGTCG) [61] and AAC-CGCTTCTACTACCAG/GCGAACAGCTCGGTCAG) [62]. The PCR amplification protocol was as follows: 94 °C for 5 min, 30 cycles of 94 °C for 30 s, 52 °C for 30 s, 72 °C for 30 s and 10 min at 72 °C. PCR amplicons were purified using the BSC48L1E-G DNA Kit (Bioer, Hangzhou, China). The purified amplicons were pooled in equimolar proportions and sequenced using the PE250 method on an Illumina Nova 6000 platform at Guangdong Magigene Biotechnology (Guangzhou, China). Sequencing data generated in this study have been deposited at the NCBI Sequence Read Archive under accession code SRP542393.

The software fastp (version 0.14.1) was used for clipping and primer removal from the raw data to obtain clean reads. Operational taxonomic units (OTUs) were defined with a 97% similarity cutoff using USEARCH. The community composition and abundance were analyzed using R microeco package (v1.12.0). Diversity indices were calculated based on OTUs using usearch-alpha_div (V10, http://www.drive5.com/usearch/ (accessed on 14 March 2024).

4.6. Biolog EcoPlates Assays of Soil Microbiota

The soil microbiota in each soil sample was analyzed using Biolog EcoPlates (Hayward, CA, USA) containing 31 different carbon sources to examine the relative carbon utilization in bacterial populations. Briefly, 10 g fresh soil was mixed with 100 mL 0.05 M pH 7 sterilized Phosphate Buffered Saline (PBS) and oscillated for 30 min. Thereafter, 1 mL of the supernatant was diluted using sterilized PBS (1×) up to 10^{-3}-fold. Subsequently, 150 μL diluted suspension was added to each well of the Biolog EcoPlate, and the microplate was incubated at 25 °C for 168 h. The optical density (OD) value of each well on the Biolog EcoPlate was measured at a wavelength of 590 nm every 12 h. Soil microbial activity was calculated as the average well-color development (AWCD), which is determined by using the following equation:

$$A = \sum(Ci - R)/31 \quad (1)$$

where Ci represents the OD value of each sample well, and R is the OD value of the control well.

4.7. Data Analysis

The data manipulation and statistical analysis were conducted using Microsoft Excel and SPSS 27.0. A one-way analysis of variance (ANOVA) with the Duncan test was performed to analyze the data, with the criterion for statistical significance set at $p < 0.05$ to determine the significant differences in physicochemical parameters among different treatments. The results were visualized using Origin 2024. Principal component analysis (PCA) was carried out using Canoco 5.0 to reveal relationships between soil properties, phosphorus fractions and microbial indicators. Pearson's correlation coefficient was employed to analyze the relationships among soil and plant variables. Heat maps were generated using ChiPlot (https://www.chiplot.online (accessed on 14 March 2024)).

5. Conclusions

Due to the limited availability of phosphorus in rice cultivation, the development of an efficient biochar-based fertilizer could achieve greater benefits. Our study indicated that soil phosphorus conversion significantly increased AP for BMB and OF combined while decreasing levels of inactive phosphorus. Simultaneously, the application of BMB and OF greatly reshaped the community structure of phosphorus-mobilizing bacteria and significantly increased microbial catabolic activity, as indicated by the highest values of AWCD. This was particularly evident in the utilization of carbohydrates, carboxylic acids, and amino acid carbons. It was confirmed that the application of BMB and OF promoted phosphorus uptake and growth in rice. Hence, this study provides a theoretical basis for achieving high yields in rice production. As well as Experimental evidence showing adsorption or desorption behavior under controlled conditions would help elucidate the role of soil particles and biochar in phosphorus dynamics.

Supplementary Materials: The following supporting information can be downloaded at: https://www.mdpi.com/article/10.3390/plants14020214/s1, Table S1. Physicochemical properties of the experimental soil, biochar and organic fertilizer. Table S2. Effects of different fertilization treatments on soil at 0, 21, 38 days. Table S3. Morphological characteristics of rice root systems at 38 days post-treatment. Figure S1. SEM and EDS images of RHB (a, b) and BMB (c, d). RHB, rick husk biochar; rice husk biochar-immobilized *Bacillus megaterium* Morphological characteristics of rice root systems at 38 days post-treatment. Figure S2. FTIR images of RHB and BMB. RHB, rick husk biochar; rice husk biochar-immobilized *Bacillus megaterium*. Figure S3. Phosphorus solubilizing capacity of BMB in different phosphorus medium. Means ± standard deviations for three replicates. Figure S4. Diversity and abundance of the *pqqC* and *phoD*-harboring bacterial community. Means ± standard deviations for three replicates. Different letters within a column indicate significant differences at $p < 0.05$. CK, control without fertilizer; OF; 0.1% organic fertilizer; RHB, 1% rice husk biochar; BOF, 1% rice husk biochar and 0.1% organic fertilizer. MOF, 1% rice husk biochar-immobilized *Bacillus megaterium* and 0.1% organic fertilizer. Figure. S5. Utilization of six major carbon source after 96 hours under different treatments. Means ± standard deviations for three replicates. Different letters within a column indicate significant differences at $p < 0.05$. CK, control without fertilizer; OF; 0.1% organic fertilizer; RHB, 1% rice husk biochar; BOF, 1% rice husk biochar and 0.1% organic fertilizer. MOF, 1% rice husk biochar-immobilized *Bacillus megaterium* and 0.1% organic fertilizer.

Author Contributions: K.Y.: Writing—original draft, Writing—review & editing, Formal analysis. Z.W.: Writing—original draft, Writing—review & editing. W.Y. and S.L.: Writing—review & editing. D.W. and H.Z.: Formal analysis. Z.Y.: Writing—review & editing, Supervision, Project administration and Conceptualization. S.Y.: Writing—review and editing, Resources, Supervision. D.L.: Resources, Funding acquisition. All authors have read and agreed to the published version of the manuscript.

Funding: This work was funded by the Key Research and Development Project of Science and Technology Department of Zhejiang Province [grant number 2022C02022 and 2023C02020], Key Technologies Research and Development Program [grant number 2023YFD1902900], and Lishui City Key R&D Program in Science and Technology [grant number 2022ZDYF16].

Data Availability Statement: Data will be made available on request.

Conflicts of Interest: The authors declare no conflict of interest.

References

1. Bora, A.; Choudhury, P.R.; Pande, V.; Mandal, A.B. Assessment of genetic purity in rice (*Oryza sativa* L.) hybrids using microsatellite markers. *3 Biotech* **2016**, *6*, 1–7. [CrossRef] [PubMed]
2. Rasul, M.; Yasmin, S.; Suleman, M.; Zaheer, A.; Reitz, T.; Tarkka, M.T.; Islam, E.; Mirza, M.S. Glucose dehydrogenase gene containing phosphobacteria for biofortification of Phosphorus with growth promotion of rice. *Microbiol. Res.* **2019**, *223–225*, 1–12. [CrossRef] [PubMed]
3. Ali, M.A.; Ajaz, M.M.; Rizwan, M.; Qayyum, M.F.; Arshad, M.; Hussain, S.; Ahmad, N.; Qureshi, M.A. Effect of biochar and phosphate solubilizing bacteria on growth and phosphorus uptake by maize in an Aridisol. *Arabian J. Geosci.* **2020**, *13*, 333. [CrossRef]
4. Timofeeva, A.; Galyamova, M.; Sedykh, S. Prospects for Using Phosphate-Solubilizing Microorganisms as Natural Fertilizers in Agriculture. *Plants* **2022**, *11*, 2119. [CrossRef]
5. Wang, T.; Liu, M.Q.; Li, H.X. Inoculation of phosphate-solubilizing bacteria Bacillus thuringiensis B1 increases available phosphorus and growth of peanut in acidic soil. *Acta Agric. Scand. Sect. B* **2014**, *64*, 252–259. [CrossRef]
6. Bai, Y.C.; Chang, Y.Y.; Hussain, M.; Lu, B.; Zhang, J.P.; Song, X.B.; Lei, X.S.; Pei, D. Soil chemical and microbiological properties are changed by long-term chemical fertilizers that limit ecosystem functioning. *Microorganisms* **2020**, *8*, 694. [CrossRef]
7. Saadouli, I.; Mosbah, A.; Ferjani, R.; Stathopoulou, P.; Galiatsatos, I.; Asimakis, E.; Marasco, R.; Daffonchio, D.; Tsiamis, G.; Ouzari, H.I. The impact of the inoculation of phosphate-solubilizing bacteria Pantoea agglomerans on phosphorus availability and bacterial community dynamics of a semi-arid soil. *Microorganisms* **2021**, *9*, 1661. [CrossRef]
8. Wang, L.; Wang, J.; Yuan, J.; Tang, Z.; Wang, J.; Zhang, Y. Long-Term Organic Fertilization Strengthens the Soil Phosphorus Cycle and Phosphorus Availability by Regulating the *pqqC*- and *phoD*-Harboring Bacterial Communities. *Microb. Ecol.* **2023**, *86*, 2716–2732. [CrossRef]
9. Choi, O.; Kim, J.; Kim, J.G.; Jeong, Y.; Moon, J.S.; Park, C.S.; Hwang, I. Pyrroloquinoline quinone is a plant growth promotion factor produced by Pseudomonas fluorescens B16. *Plant Physiol.* **2008**, *146*, 657. [CrossRef]
10. Richardson, A.E.; Simpson, R.J. Soil microorganisms mediating phosphorus availability update on microbial phosphorus. *Plant Physiol.* **2011**, *156*, 989–996. [CrossRef]
11. Jiang, Y.; Yang, F.; Dai, M.; Ali, I.; Shen, X.; Hou, X.; Alhewairini, S.S.; Peng, C.; Naz, I. Application of microbial immobilization technology for remediation of Cr(VI) contamination: A review. *Chemosphere* **2022**, *286*, 131721. [CrossRef] [PubMed]
12. Zheng, B.X.; Ding, K.; Yang, X.R.; Wadaan, M.A.M.; Hozzein, W.N.; Peñuelas, J.; Zhu, Y.G. Straw biochar increases the abundance of inorganic phosphate solubilizing bacterial community for better rape (*Brassica napus*) growth and phosphate uptake. *Sci. Total Environ.* **2019**, *647*, 1113–1120. [CrossRef] [PubMed]
13. Herrmann, L.; Lesueur, D.; Robin, A.; Robain, H.; Wiriyakitnateekul, W.; Bräu, L. Impact of biochar application dose on soil microbial communities associated with rubber trees in North East Thailand. *Sci. Total Environ.* **2019**, *689*, 970–979. [CrossRef]
14. Gul, S.; Whalen, J.K. Biochemical cycling of nitrogen and phosphorus in biochar-amended soils. *Soil Biol. Biochem.* **2016**, *103*, 1–15. [CrossRef]
15. Husna, N.; Budianta, D.; Munandar, M.; Napoleon, A. Evaluation of Several Biochar Types as Inoculant Carrier for Indigenous Phosphate Solubilizing Microoorganism from Acid Sulphate Soil. *J. Ecol. Eng.* **2019**, *20*, 1–8. [CrossRef]
16. Lehmann, J.; Rillig, M.C.; Thies, J.; Masiello, C.A.; Hockaday, W.C.; Crowley, D. Biochar effects on soil biota–a review. *Soil Biol. Biochem.* **2011**, *43*, 1812–1836. [CrossRef]
17. Lu, J.; Liu, S.; Chen, W.; Meng, J. Study on the mechanism of biochar affecting the effectiveness of phosphate solubilizing bacteria. *World J. Microbiol. Biotechnol.* **2023**, *39*, 87. [CrossRef]
18. Azeem, M.; Hassan, T.U.; Tahir, M.I.; Ali, A.; Jeyasundar, P.G.S.A.; Hussain, Q.; Bashir, S.; Mehmood, S.; Zhang, Z. Tea leaves biochar as a carrier of Bacillus cereus improves the soil function and crop productivity. *Appl. Soil Ecol.* **2021**, *157*, 103732. [CrossRef]
19. Luo, Z.; Yao, B.; Yang, X.; Wang, L.; Xu, Z.; Yan, X.; Tian, L.; Zhou, H.; Zhou, Y. Novel insights into the adsorption of organic contaminants by biochar: A review. *Chemosphere* **2022**, *287*, 132113. [CrossRef]
20. Pandian, K.; Samiyappan, K.; Rangasamy, A.; Raju, I.; Chitraputhirapillai, S.; Bose, J. Bacillus megaterium-Embedded Organo Biochar Phosphorous Fertilizer Improves Soil Microbiome and Nutrient Availability to Enhance Black Gram (*Vigna mungo* L) Growth and Yield. *J. Soil Sci. Plant Nutr.* **2024**. [CrossRef]
21. Jiang, Z.; Li, Q.; Peng, F.; Yu, J. Biochar Loaded with a Bacterial Strain N33 Facilitates Pecan Seedling Growth and Shapes Rhizosphere Microbial Community. *Plants* **2024**, *13*, 1226. [CrossRef] [PubMed]
22. Bai, K.; Wang, W.; Zhang, J.; Yao, P.; Cai, C.; Xie, Z.; Luo, L.; Li, T.; Wang, Z. Effects of phosphorus-solubilizing bacteria and biochar application on phosphorus availability and tomato growth under phosphorus stress. *BMC Biol.* **2024**, *22*, 211. [CrossRef] [PubMed]

23. Chen, W.; Wu, Z.; Liu, C.; Zhang, Z.; Liu, X. Biochar combined with Bacillus subtilis SL-44 as an eco-friendly strategy to improve soil fertility, reduce Fusarium wilt, and promote radish growth. *Ecotoxicol. Environ. Saf.* **2023**, *251*, 114509. [CrossRef] [PubMed]
24. Chen, H.; Zhang, J.; Tang, L.; Su, M.; Tian, D.; Zhang, L.; Li, Z.; Hu, S. Enhanced Pb immobilization via the combination of biochar and phosphate solubilizing bacteria. *Environ. Int.* **2019**, *127*, 395–401. [CrossRef]
25. Alkahtani, J.; Elshikh, M.S.; Alwahibi, M.S.; Muhammad, A.; Khalid, S. Phosphorus and Zinc Fertilization Improve Productivity and Profitability of Rice Cultivars under Rice-Wheat System. *Agronomy* **2020**, *10*, 1085. [CrossRef]
26. Bi, Q.F.; Li, K.J.; Zheng, B.X.; Liu, X.P.; Li, H.Z.; Jin, B.J.; Ding, K.; Yang, X.R.; Lin, X.Y.; Zhu, Y.G. Partial replacement of inorganic phosphorus (P) by organic manure reshapes phosphate mobilizing bacterial community and promotes P bioavailability in a paddy soil. *Sci. Total Environ.* **2020**, *703*, 134977. [CrossRef]
27. Hu, W.; Zhang, Y.; Rong, X.; Zhou, X.; Fei, J.; Peng, J.; Luo, G. Biochar and organic fertilizer applications enhance soil functional microbial abundance and agroecosystem multifunctionality. *Biochar* **2024**, *6*, 3. [CrossRef]
28. Tu, C.; Wei, J.; Guan, F.; Liu, Y.; Sun, Y.; Luo, Y. Biochar and bacteria inoculated biochar enhanced Cd and Cu immobilization and enzymatic activity in a polluted soil. *Environ. Int.* **2020**, *137*, 105576. [CrossRef]
29. Ducousso-Détrez, A.; Fontaine, J.; Lounès-Hadj Sahraoui, A.; Hijri, M. Diversity of phosphate chemical forms in soils and their contributions on soil microbial community structure changes. *Microorganisms* **2022**, *10*, 609. [CrossRef]
30. Yan, Z.; Chen, S.; Li, J.; Alva, A.; Chen, Q. Manure and nitrogen application enhances soil phosphorus mobility in calcareous soil in greenhouses. *J. Environ. Manag.* **2016**, *181*, 26–35. [CrossRef]
31. Fan, Y.; Lv, G.; Chen, Y.; Chang, Y.; Li, Z. Differential effects of cow dung and its biochar on Populus euphratica soil phosphorus effectiveness, bacterial community diversity and functional genes for phosphorus conversion. *Front. Plant Sci.* **2023**, *14*, 1242469. [CrossRef] [PubMed]
32. Richards, S.; Marshall, R.; Lag-Brotons, A.J.; Semple, K.T.; Stutter, M. Phosphorus solubility changes following additions of bioenergy wastes to an agricultural soil: Implications for crop availability and environmental mobility. *Geoderma* **2021**, *401*, 115150. [CrossRef]
33. Fujita, K.; Kunito, T.; Moro, H.; Toda, H.; Otsuka, S.; Nagaoka, K. Microbial resource allocation for phosphatase synthesis reflects the availability of inorganic phosphorus across various soils. *Biogeochemistry* **2017**, *136*, 325–339. [CrossRef]
34. Sui, L.; Tang, C.; Cheng, K.; Yang, F. Biochar addition regulates soil phosphorus fractions and improves release of available phosphorus under freezing-thawing cycles. *Sci. Total Environ.* **2022**, *848*, 157748. [CrossRef] [PubMed]
35. Li, R.; Zhang, S.; Zhang, M.; Fei, C.; Ding, X. Phosphorus fractions and adsorption–desorption in aggregates in coastal saline-alkaline paddy soil with organic fertilizer application. *J. Soils Sediments* **2021**, *21*, 3084–3097. [CrossRef]
36. Wang, C.; Xue, L.; Jiao, R. Soil phosphorus fractions, phosphatase activity, and the abundance of phoC and phoD genes vary with planting density in subtropical Chinese fir plantations. *Soil Tillage Res.* **2021**, *209*, 104946. [CrossRef]
37. Zhang, Y.; Li, Y.; Wang, S.; Umbreen, S.; Zhou, C. Soil phosphorus fractionation and its association with soil phosphate-solubilizing bacteria in a chronosequence of vegetation restoration. *Ecol. Eng.* **2021**, *164*, 106208. [CrossRef]
38. Qi, W.-Y.; Chen, H.; Wang, Z.; Xing, S.-F.; Song, C.; Yan, Z.; Wang, S.-G. Biochar-immobilized Bacillus megaterium enhances Cd immobilization in soil and promotes Brassica chinensis growth. *J. Hazard. Mater.* **2023**, *458*, 131921. [CrossRef]
39. Macias-Benitez, S.; Garcia-Martinez, A.M.; Caballero Jimenez, P.; Gonzalez, J.M.; Tejada Moral, M.; Parrado Rubio, J. Rhizospheric organic acids as biostimulants: Monitoring feedbacks on soil microorganisms and biochemical properties. *Front. Plant Sci.* **2020**, *11*, 633. [CrossRef]
40. Li, N.; Sheng, K.; Zheng, Q.; Hu, D.; Zhang, L.; Wang, J.; Zhang, W. Inoculation with phosphate-solubilizing bacteria alters microbial community and activates soil phosphorus supply to promote maize growth. *Land Degrad. Dev.* **2023**, *34*, 777–788. [CrossRef]
41. Nguyen, T.T.N.; Wallace, H.M.; Xu, C.Y.; Zwieten, L.; Weng, Z.H.; Xu, Z.; Che, R.; Tahmasbian, I.; Hu, H.W.; Bai, S.H. The effects of short term, long term and reapplication of biochar on soil bacteria. *Sci. Total Environ.* **2018**, *636*, 142–151. [CrossRef] [PubMed]
42. Chen, X.; Jiang, N.; Condron, L.M.; Dunfield, K.E.; Chen, Z.; Wang, J.; Chen, L. Soil alkaline phosphatase activity and bacterial phoD gene abundance and diversity under long-term nitrogen and manure inputs. *Geoderma* **2019**, *349*, 36–44. [CrossRef]
43. Liu, J.; Ma, Q.; Hui, X.; Ran, J.; Ma, Q.; Wang, X.; Wang, Z. Long-term high-P fertilizer input decreased the total bacterial diversity but not phoD-harboring bacteria in wheat rhizosphere soil with available-P deficiency. *Soil Biol. Biochem.* **2020**, *149*, 107918. [CrossRef]
44. Dai, Z.; Liu, G.; Chen, H.; Chen, C.; Wang, J.; Ai, S.; Wei, D.; Li, D.; Ma, B.; Tang, C. Long-term nutrient inputs shift soil microbial functional profiles of phosphorus cycling in diverse agroecosystems. *ISME J.* **2020**, *14*, 757–770. [CrossRef]
45. Ikoyi, I.; Fowler, A.; Schmalenberger, A. One-time phosphate fertilizer application to grassland columns modifies the soil microbiota and limits its role in ecosystem services. *Sci. Total Environ.* **2018**, *630*, 849–858. [CrossRef]

46. Huang, Z.; Qin, Y.; He, X.; Zhang, M.; Ren, X.; Yu, W.; Ji, K. Analysis on metabolic functions of rhizosphere microbial communities of Pinus massoniana provenances with different carbon storage by Biolog Eco microplates. *Front. Microbiol.* **2024**, *15*, 1365111. [CrossRef]
47. Klimek, B.; Chodak, M.; Jaźwa, M.; Niklińska, M. Functional diversity of soil microbial communities in boreal and temperate Scots pine forests. *Eur. J. For. Res.* **2016**, *135*, 731–742. [CrossRef]
48. Liu, X.; Liu, H.; Zhang, Y.; Chen, G.; Li, Z.; Zhang, M. Straw return drives soil microbial community assemblage to change metabolic processes for soil quality amendment in a rice-wheat rotation system. *Soil Biol. Biochem.* **2023**, *185*, 109131. [CrossRef]
49. Tian, J.; Wang, J.; Dippold, M.; Gao, Y.; Blagodatskaya, E.; Kuzyakov, Y. Biochar affects soil organic matter cycling and microbial functions but does not alter microbial community structure in a paddy soil. *Sci. Total Environ.* **2016**, *556*, 89–97. [CrossRef]
50. Ren, H.; Li, Z.; Chen, H.; Zhou, J.; Lv, C. Effects of Biochar and Plant Growth-Promoting Rhizobacteria on Plant Performance and Soil Environmental Stability. *Sustainability* **2022**, *14*, 10922. [CrossRef]
51. Tan, L.; Zhou, L.; Zhao, N.; He, Y.; Qiu, Z. Development of a low-cost portable device for pixel-wise leaf SPAD estimation and blade-level SPAD distribution visualization using color sensing. *Comput. Electron. Agric.* **2021**, *190*, 106487. [CrossRef]
52. Thula, S.; Moturu, T.R.; Salava, H.; Balakhonova, V.; Berka, M.; Kerchev, P.; Mishra, K.B.; Nodzynski, T.; Simon, S. Strigolactones stimulate high light stress adaptation by modulating photosynthesis rate in Arabidopsis. *J. Plant Growth Regul.* **2023**, *42*, 4818–4833. [CrossRef]
53. Wu, T.; Li, C.; Xing, X.; Pan, X.; Liu, C.; Tian, Y.; Wang, Z.; Zhao, J.; Wang, J.; He, B. Straw return and organic fertilizers instead of chemical fertilizers on growth, yield and quality of rice. *Earth. Sci. Inform.* **2022**, *15*, 1363–1369. [CrossRef]
54. He, L.; Zhong, H.; Liu, G.; Dai, Z.; Brookes, P.C.; Xu, J. Remediation of heavy metal contaminated soils by biochar: Mechanisms, potential risks and applications in China. *Environ. Pollut.* **2019**, *252*, 846–855. [CrossRef]
55. Siddiq, S.; Saleem, U.; Ahmad, K.; Anayat, A.; Affan, Q.M.; Anwar, M.F.; Nazir, H.; Asghar, N. Comparison of conventional and non-conventional carriers for bacterial survival and plant growth. *Int. J. Agric. Innov. Res.* **2018**, *6*, 126–129. [CrossRef]
56. Naureen, Z.; Rehman, N.U.; Hussain, H.; Gilani, S.A.; Khan, A.L.; Harrasi, A.A. Exploring the potentials of Lysinibacillus sphaericus ZA9 for plant growth promotion and biocontrol activities against phytopathogenic fungi. *Front. Microbiol.* **2017**, *8*, 252103. [CrossRef]
57. Zhu, X.; Li, X.; Shen, B.; Zhang, Z.; Wang, J.; Shang, X. Bioremediation of lead-contaminated soil by inorganic phosphate-solubilizing bacteria immobilized on biochar. *Ecotoxicol. Environ. Saf.* **2022**, *237*, 113524. [CrossRef]
58. Lu, R.K. *Analytical Methods of Soil Agricultural Chemistry*; China Agricultural Science and Technology Press: Beijing, China, 1999.
59. Tabatabai, M.A.; Bremner, J.M. Use of p-nitrophenyl phosphate for assay of soil phosphatase activity. *Soil Biol. Biochem.* **1969**, *1*, 301–307. [CrossRef]
60. Tiessen, H.; Moir, J.O. Characterization of available P by sequential extraction. In *Soil Sampling and Methods of Analysis*; Lewis Publishers: Boca Raton, FL, USA, 1993.
61. Sakurai, M.; Wasaki, J.; Tomizawa, Y.; Shinano, T.; Osaki, M. Analysis of bacterial communities on alkaline phosphatase genes in soil supplied with organic matter. *Soil Sci. Plant Nutr.* **2008**, *54*, 62–71. [CrossRef]
62. Zheng, B.X.; Hao, X.L.; Ding, K.; Zhou, G.W.; Chen, Q.L.; Zhang, J.B.; Zhu, Y.G. Long-term nitrogen fertilization decreased the abundance of inorganic phosphate solubilizing bacteria in an alkaline soil. *Sci. Rep.* **2017**, *7*, 42284. [CrossRef]

plants

MDPI

Article

Yield and Sensorial and Nutritional Quality of Strawberry (*Fragaria × ananassa* Duch.) Fruits from Plants Grown Under Different Amounts of Irrigation in Soilless Cultivation

Davide Raffaelli, Rohullah Qaderi, Luca Mazzoni, Bruno Mezzetti and Franco Capocasa *

Department of Agricultural, Food and Environmental Sciences, Università Politecnica delle Marche (UNIVPM), Via Brecce Bianche 10, 60131 Ancona, Italy; d.raffaelli@pm.univpm.it (D.R.); r.qaderi@staff.univpm.it (R.Q.); l.mazzoni@staff.univpm.it (L.M.); b.mezzetti@staff.univpm.it (B.M.)

* Correspondence: f.capocasa@staff.univpm.it

Abstract: Water scarcity is an ecological issue affecting over 10% of Europe. It is intensified by rising temperatures, leading to greater evaporation and reduced precipitation. Agriculture has been confirmed as the sector accounting for the highest water consumption globally, and it faces significant challenges relating to drought, impacting crop yields and food security. Sustainable practices, precision irrigation, and the development of drought-resistant crops are essential for the mitigation of this threat. Effective, innovative solutions are crucial for optimizing water use for intensive crops such as cultivated strawberries (*Fragaria × ananassa*). This study emphasizes the importance of identifying the genotypes most resilient to low water availability. Experimental trials involving reduced irrigation levels were set up to identify genotypes with a greater capacity to increase fruit quality and maintain fruit yield. Reduced water conditions positively influenced strawberry fruit quality, exhibiting improved citric acid, soluble solids, and color brightness linked to decreased water use, while firmness remained stable. Notably, the total phenolic content was most affected by stress, indicating strong antioxidant responses. With these interesting variations in fruit quality came a different response in plant yield. Plants belonging to the Lauretta and AN15,07,53 cultivars maintained a 98% fruit yield when grown under WS1 conditions. While the yield for the Francesca cultivar increased by 10% under the stressed WS1 conditions in comparison to the control conditions, water stress in the WS2 treatment caused a strong reduction in yield in all three genotypes. Overall, the findings emphasize the importance of identifying for each new cultivar the most appropriate water regime in order to amplify the quality of the fruit, thus maintaining high production standards and saving water.

Keywords: *Fragaria × ananassa*; strawberry; soilless; nutritional quality; yield; water management

1. Introduction

The biggest ecological issues are currently water availability and distribution [1]. Many regions of the world are facing critical situations due to water scarcity, exacerbated by rising temperatures that increase evaporation rates and consequently raise the demand for water for crops [2]. Water scarcity affects more than 10% of Europe, particularly the southern regions, such as Italy and Spain [3]. Agriculture is the largest consumer of water compared to other sectors, accounting for 70% of the total annual water withdrawal [4,5]. Drought is an abiotic stress that hampers normal plant physiological activity, leading to physiological changes in plant structure, such as stomatal closure, reduced photosynthesis, minimized evapotranspiration, reduced biomass, and, consequently, decreased fruit production [6]. In this scenario, drought conditions are expected to worsen due to climate change, with rising air temperatures and increased atmospheric CO_2 levels altering rainfall patterns and distribution [7–10]. Due to this issue, numerous experiments on precision irrigation and the selection of the most drought-resistant plants have been conducted for many crop species worldwide [11,12]. Among the tested crops, berry plants have gained attention due to the rising interest in them on the market [13,14]. Among berry plants, strawberries (*Fragaria* × *ananassa*) have been observed to have the highest demand, with 95.6×10^3 tons produced globally in 2022 [15]. Strawberries contain high concentrations of molecules with antioxidant activity, such as vitamins, polyphenols, and phenolic acids, qualifying them as bio-functional foods [16]. Genetic diversity has been the basis for strawberry improvement because it is still an essential source of important traits that allow plants to adapt to climatic changes. In general, crops' wild relatives exhibit valuable traits not maintained among cultivars, and these traits are critical for increasing adaptation to diverse environmental conditions [17]. Strawberry backcrossing programs of *F.* × *ananassa* combined with the two wild strawberry species (*F. virginiana* or *F. chiloensis)* are fundamental for generating new pre-breeding material carrying new essential traits [18]. Genomic studies on different breeding populations are now making progress in identifying adaptive and horticultural traits of high importance for genetically improving strawberry plants through the identification of new resilient cultivars [19,20]. Numerous studies have highlighted the critical need to investigate water-saving methods to optimize resource use [21,22] and examine changes in fruit quality. In strawberry cultivation, it is crucial to define the most appropriate irrigation regime, depending on the cultivar, the environment, and the cultivation system, based on the concept of returning evapotranspiration water from the plant. Determining the precise amount of water consumed by strawberries is challenging due to significant variations caused by genotypes and environmental conditions. However, earlier research [23] reported that in soil-based farming systems in the southeastern Marche region, water usage averages around 1120 m^3/ha under standard conditions, with a water use efficiency (WUE) of approximately 37%. Tracking water used in conventional farming systems has proven to be complex, as precise measurements are difficult to maintain. Conversely, soilless farming systems are considered more adaptable, allowing for optimized resource management throughout the cultivation process. According to [24], WUE in soilless cultivation typically ranges between 31% and 41%, reflecting an improvement in water use flexibility and efficiency compared to traditional methods. The controlled water deficit approach is currently considered essential in order to be able to guarantee the correct use of water, with possible savings, maintaining yield and increasing quality. Marcellini et al. [23] revealed a significant gap in farmers' knowledge of irrigation management, suggesting that water can be conserved without compromising production or quality. However, few studies have described the changes in bio-functional compound composition in strawberries from plants subjected to reduced irrigation. Our two-year study aimed, firstly, to investigate the productive, qualitative, and nutritional profiles of fruits harvested from strawberry genotypes of the UNIVPM breeding program; evaluate how the tested plants responded to different irrigation rates; and identify the most performant genotypes to use as parental materials. Secondarily, we tried to better understand how water shortages influence the bio-functional molecules accumulated in the fruits during cultivation and how dietary antioxidant compounds are correlated with each other.

2. Results and Discussion

2.1. Environmental Parameter

The water consumption of the plants was taken into consideration for two autumn production cycles: from 12 September to 7 November 2022, and from 9th September to 4th November 2023. The data are expressed in % of water volume content (%wvc), as shown in Figure 1. The plants at treatment WW maintained wvc around 44%, those at WS1 maintained wvc around 30%, and those at WS2 reached a level of humidity below 20%. Finally, the total water consumption reached 216 L for WW (100%), 160 L for WS1 (75%), and 130 L (60%).

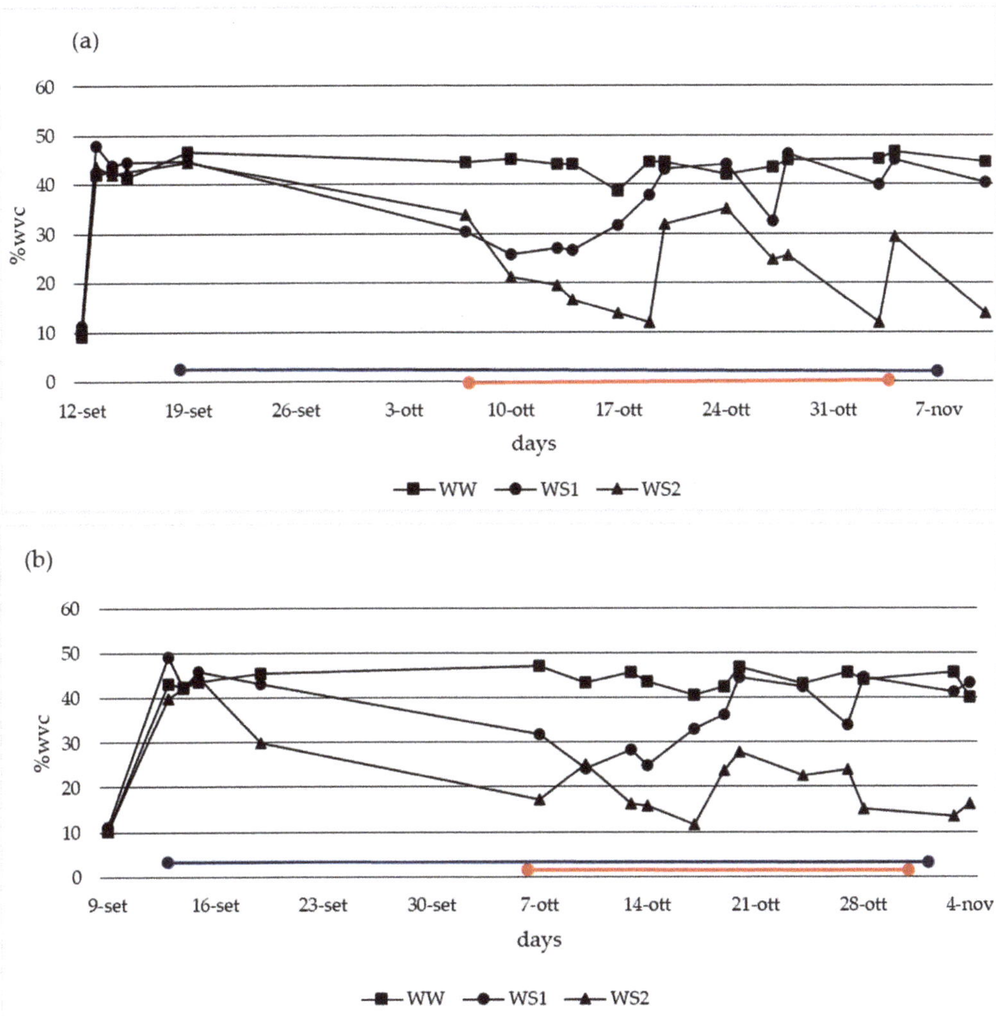

Figure 1. Water volume content (%wvc) of the three different treatments in 2022 (**a**) and 2023 (**b**). Blue line represents diversified irrigation periods; red line represents harvesting period.

2.2. Productive Parameters

We assessed the plants' marketable yield (MY), the average fruit weight (AFW), and water use efficiency (WUE). Both the genotype and the treatment showed a crucial influence on fruit yield/plant and average fruit weight (Table 1), as investigated in other experiments [23]. At full irrigation (WW), the selection AN15,07,53 displayed the highest MY (185.52 g/plant), followed by 'Lauretta' (166.04 g/plant) and 'Francesca' (91.66 g/plant), as shown in Table 2. At WS1 for all genotypes, the MY was not statistically different from the WW treatment. For example, AN15,07,53 reached a marketable yield of 181.76 g/plant, 'Lauretta' reached 164.09 g/plant, and 'Francesca' reached 100.33 g/plant. The critical lack of production at WS2 seemed to be too high in comparison to the other treatments, with 91.33 g/plant for AN15,07,53, 85.47 g/plant for 'Lauretta', and 56.9 g/plant for 'Francesca'. The investigation by Ödemiş et al., 2020 [25], revealed a similar trend. Gao et al., 2021 [26], explained the lack of production by the decrease in the plant's photosynthetic activity. Regarding the AFW at full irrigation, the cultivar 'Francesca' showed the largest size, 14.11 g/fruit, while 'Lauretta' and the selection AN15,07,53 reached, respectively, 10.93 g/fruit and 10.09 g/fruit. Similarly to the loss of MY, the WS2 treatment exhibited the lowest values regarding the fruit weight, with 6.61 g/fruit for AN15,07,53, 8.77 g/fruit for 'Lauretta', and 8.1 g/fruit for 'Francesca'. WUE had a remarkable role, and it was obtained from the ratio between kg of fruit production and the m^3 of water supplied through irrigation. The WW did not prove to be the most efficient treatment, with a lack of efficiency in comparison with WS1 of about 25% in AN15,07,53, 33% in 'Francesca', and 26% in 'Lauretta'. WS1 turned out to be the most efficient irrigation. The genotype AN15,07,53 was revealed to be the most efficient, producing 34 kg/m^3; followed by 'Lauretta', with 30 kg/m^3 of fruits; and 'Francesca', with 18.81 kg/m^3 of fruits. WS2 had the most inefficient irrigation, reaching a loss of fruit production of 39% in AN15,07,53, 31% in 'Lauretta', and 36% in 'Francesca'.

Table 1. Three-way analysis of variance (ANOVA) for marketable yield per plant (MY) and average fruit weight (AFW). The interaction among parameters was considered strongly significant with p values < 0.01 (**), slightly significant with p values < 0.05 (*), and non-significant with p values > 0.05 (NS).

Parameters	MY	AFW
Genotype	*	**
Year	NS	NS
Treatments	**	**
Genotype × Year	NS	NS
Genotype × Treatments	NS	NS
Year × Treatments	NS	NS
Genotype × Year × Treatments	NS	NS

Table 2. Two-way ANOVA effect of different irrigation levels on the mean values (2022–2023) of marketable yield per plant (MY), average fruit weight (AFW) ± standard error is reported, and water use efficiency (WUE). Values with the same lowercase letter for the same parameter were not statistically different for Fisher's LSD test (p < 0.05).

Treatment	Genotype	MY (g/Plant)	AFW (g/Fruit)	WUE (kg MY/m^3 Water)
WW	AN15,07,53	185.52 ± 10.5 a	10.09 ± 2.68 bc	25.76
WS1	AN15,07,53	181.76 ± 9.67 a	8.48 ± 3.32 cd	34.08
WS2	AN15,07,53	91.33 ± 4.65 bcd	6.61 ± 2.6 e	21.07
WW	Francesca	91.66 ± 5.93 bcd	14.11 ± 3.48 a	12.73

Table 2. *Cont.*

Treatment	Genotype	MY (g/Plant)	AFW (g/Fruit)	WUE (kg MY/m^3 Water)
WS1	Francesca	100.33 ± 8.11 bc	12.05 ± 4.66 ab	18.811
WS2	Francesca	56.9 ± 8.17 d	8.77 ± 4.19 cd	13.13
WW	Lauretta	166.04 ± 7.01 ab	10.93 ± 2.96 bc	23.06
WS1	Lauretta	164.09 ± 6.55 ab	9.41 ± 3.1 bcd	30.766
WS2	Lauretta	85.47 ± 7.08 bcd	8.1 ± 3.07 d	19.72

2.3. Sensorial Quality

The sensory quality of strawberry fruit is crucial, as it is directly perceived by the consumer and influences the commercial value of these fruits. Organoleptic quality is typically evaluated based on key parameters such as appearance, flavor, and firmness, which are indicators of fruit ripeness. These characteristics are primarily influenced by genotype and cultivation conditions. Both the genotype and the treatment broadly contribute to the expression of the fruit's color, firmness, °Brix, and titratable acidity (Table 3). Specifically (Table 4), at WW AN15,07,53 showed the highest brightness (L*) in comparison to the other genotypes, with an average value of 40.16 L* compared to 'Lauretta' (37.42 L*) and 'Francesca' (36.27 L*). Moreover, the Chroma value, which is derived from the color dimension coordinates 'a' and 'b' using the formula $[(a^2 + b^2)]^{1/2}$ to measure color saturation, was included as part of the organoleptic evaluation. It showed no visible differences in AN15,07,53 and 'Francesca' at full irrigation, with, respectively, 51.33 and 53.46 values. On the contrary, Lauretta' exhibited a minor intensity in all the treatments. Concerning the fruits' firmness, the results did not prove high differences between the genotypes at WW; in fact, AN15,07,53 reached 456.36 g/cm^2, 'Francesca' reached 433.63 g/cm^2, and 'Lauretta' reached 422.72 g/cm^2. Concerning the soluble solids (SS) of the fruits, in WW conditions, the AN15,07,53 genotypes exhibited 7.3 °Brix, 'Francesca' exhibited 7.6 °Brix, and 'Lauretta' exhibited 7 °Brix. The citric acid (CA) at WW was estimated at around 0.63% for AN15,07,53, 0.75% for 'Francesca', and 0.7% for 'Lauretta'. WS2 resulted in a rise in fruit acidity. Comparing the treatments, we can conclude the following considerations. The reduction in water to WS1 and, more evidently, to WS2 implied a rise in the final fruit brightness in all genotypes. This may be attributed to the reduced canopy of the plants under water shortages, which likely leads to decreased light penetration within the canopy and subsequently affects the pigmentation of the fruits. [27,28]. For example, concerning the chroma values, a noticeable distinction was evident between WW and WS2 (e.g., the chroma of 'Lauretta' fruits at WW was 47.49, instead of at WS2, where it reached 48.35). Regarding the firmness of the fruits, the shortening of the irrigation volumes implied a hardening of the fruit consistency (e.g., in the WS1 condition, AN15,07,53 reached 473.63 g/cm^2, 'Francesca' reached 470.9 g/cm^2, and 'Lauretta' reached 455.45 g/cm^2, from a consistency at WW of 456.36 g/cm^2 for AN,15,07,53, 433.63 g/cm^2 for 'Francesca', and 422.72 g/cm^2 for 'Lauretta'. A notable observation was the flattening of sweetness across all genotypes at WW, along with a general tendency to increase the °Brix content in the fruits from plants treated under WS2 conditions. Many studies revealed a similar tendency, entailing the sugar concentration as a response to drought conditions [10,16,29]. Some stimulating considerations may be given to the relationship between SS and drought tolerance since there is a positive correlation between the irrigation supply and the accumulation of soluble solids in many studies. The relation between the soluble solids concentration and the susceptibility to biotic stressors was positively described [30] concerning the *Botrytis cinerea* infection. Some considerations about the abiotic stressors and SSC of fruits may be interesting for future investigations. CA also exhibited correspondence between increased acidity and water reduction, as noticed in other studies [10,31]. As well as for the SS, the trend of the CA concentration suggests that different irrigation supplies may lead to different variations in respiratory metabolisms and biochemical pathways in citric acid regulation. In fact, in our study, the differences between WW and WS1 are similar; however, under WS2 conditions, the 'Francesca' plants showed a higher concentration of acidity in the fruits (e.g., WS2 in AN15,07,53 was about 0.75%; in 'Francesca', it was about 1.01%; and in 'Lauretta', it was about 0.96%). In contrast, other studies [32] demonstrated opposite behaviors for the citric acid concentration, showing a reduction in fruit acidity in plants subjected to drought conditions. In conclusion, at WS2, the plants exhibited brighter, sweeter, and sharper fruits but with irreversibly compromised fruit yield. The WS1 treatment resulted in a slight improvement in the fruit's sweetness and acidity without compromising yield. Therefore, WS1 enabled 25% savings of water for irrigation, maintaining an appreciable fruit quality.

Table 3. Three-way analysis of variance (ANOVA) for brightness (L*), chroma, firmness (F), soluble solids (SS), and citric acid content (CA). The interaction among parameters was considered strongly significant, with p values < 0.01 (**), slightly significant with p values < 0.05 (*), and non-significant with p values > 0.05 (NS).

Parameters	L*	Chroma	F	SS	CA
Genotype	**	**	**	*	**
Year	NS	NS	NS	NS	NS
Treatments	**	**	**	*	**
Genotype × Year	NS	NS	NS	NS	NS
Genotype × Treatments	**	**	NS	NS	NS
Year × Treatments	NS	NS	NS	NS	NS
Genotype × Year × Treatments	NS	NS	NS	NS	NS

Table 4. Two-way ANOVA effect of different irrigation levels on the two years mean values (2022–2023) of brightness (L*), chroma, soluble solids (SS), firmness (F), and citric acid content (CA) ± standard error (2022–2023) are reported. Values with the same lowercase letter for the same parameter were not statistically different for Fisher's LSD test ($p < 0.05$).

Treatment	Genotype	L*	Chroma	F	°Brix	CA
WW	AN15.07,53	40.16 ± 0.36 bc	51.33 ± 0.48 c	456.36 ± 3.02 ab	7.3 ± 0.27 bc	0.63 ± 0.032 e
WS1	AN15,07,53	40.52 ± 0.38 ab	52.03 ± 0.40 bc	473.63 ± 2.44 a	7.6 ± 0.54 bc	0.62 ± 0.01 e
WS2	AN15,07,53	41.18 ± 0.35 a	52.37 ± 0.54 bc	488.18 ± 2.81 a	7.9 ± 0.64 bc	0.75 ± 0.03 d
WW	Francesca	36.27 ± 0.26 h	53.46 ± 0.42 bc	433.63 ± 1.53 b	7.6 ± 0.14 bc	0.75 ± 0.05 d
WS1	Francesca	38.57 ± 0.33 de	53.51 ± 0.21 bc	470.9 ± 1.79 a	8.3 ± 0.27 ab	0.78 ± 0.03 cd
WS2	Francesca	39.47 ± 0.37 cd	55.29 ± 0.30 a	471.81 ± 1.61 a	9.1 ± 0.21 a	1.01 ± 0.03 a
WW	Lauretta	37.42 ± 0.28 fg	47.49 ± 0.37 e	422.72 ± 1.68 b	7 ± 0.69 c	0.79 ± 0.02 cd
WS1	Lauretta	37.79 ± 0.25 ef	49.68 ± 0.30 d	454.54 ± 3.02 ab	7.5 ± 0.23 bc	0.88 ± 0.03 bc
WS2	Lauretta	38.87 ± 0.29 de	48.35 ± 0.32 de	455.45 ± 2.7 ab	8 ± 0.24 bc	0.96 ± 0.02 ab

2.4. Nutritional Quality

Behind the scale-up of berry fruit consumption in recent decades, there is the consumer's perception of the healthiness provided by the habitual consumption of these fruits, and generally by vegetable food products, as part of a common diet [33–37]. The antioxidant action provided by the bio-functional molecules present in strawberry fruits demonstrates effective radical scavenging activity against reactive oxygen species (ROS). [38,39]. The key molecules are vitamins, such as ascorbic acid (Asc.A.), folate (Fol.), flavonoids, and mainly anthocyanins, with the preponderant presence of perlargonidine-3-glucoside (Pel-3-gluc), cyanidine 3-glucoside (Cya.3-gluc), pelargonidine-3-rutinoside (Pel.3-rut), and phenolic compounds, such as chlorogenic acid (Chl.A.), ellagic acid (Ell.A.), and caffeic acid (Caff.A.) [38, 40]. As highlighted by the experimentation (Table 5), the genotype has a consistent role in defining all the bio-functional molecules [41,42]. The irrigation treatment, too, showed great influence on the appearance of the following compounds, except for chlorogenic acid and caffeic acid, which appeared to be more correlated to the cultivar than to the treatment. Specifically, the genotype had a relevant influence on all the analyzed parameters (Table 6). In detail, under WW conditions, the cultivar 'Lauretta' exhibited the highest polyphenol cumulation (TPH), with107.7 mg GA/100 g; followed by 'Francesca', with 103.8 mg GA/100 g; and AN15,07,53 with 80.1 mg GA/100 g. Concerning the fruits' total antioxidant capacity (TAC), 'Lauretta' exhibited the best performance, with 334.3 mg TROLOX eq/100 g; followed by 'Francesca', with 297.6 mg TROLOX eq/100 g; and AN15,07,53, at 252.01 mg TROLOX eq/100 g. Under WW conditions, the anthocyanin compounds (ACY) were most prevalent in 'Lauretta' with, respectively, 21.5 mg of Pel-3-gluc, 0.75 mg of Cya-gluc, and 1.97 mg of Pel-3-rut. Concerning the Pel-3-gluc, AN15,07,53 and 'Francesca', exhibited remarkable values with, respectively, 18 mg/100 g and 16 mg/100 g. Interestingly, the values of the Asc.A. content in 'Francesca' and 'Lauretta' with WW irrigation reached, respectively, 32.38 mg and 30.19 mg, in contrast with the selection, which reached 20.1 mg/100 g. A similar trend was observed in the FOL fruit content, where the cultivars exhibited the highest levels, with approximately 38.4 µg/100 g for 'Francesca' and 30.1 µg/100 g for 'Lauretta,' while the fruit of AN15, 07, 53 contained 28.6 µg/100 g. Finally, Chl.A. and Ell.A. are the most common phenolic acids present in fruits. 'Francesca' proved to be the most performant, with 16.7 mg (Chl.A.) and 9.1 mg (Ell.A.). 'Lauretta' and AN15,07,53 showed similar values, with, respectively, 9.3 mg/100 g and 9.4 mg/100 g for Chl.A. and 8.6 mg/100 g and 8.5 mg/100 g for Ell.A. The reduced irrigation entailed some changes in the fruits' compound accumulation. The TPH analysis demonstrated a general trend across all genotypes, showing an increase in polyphenol fruit content with reduced water restitution (e.g., 'Lauretta' from 107.7 mg/100 g at WW to 122 mg/100 g at WS2, 'Francesca' from 103.5 mg/100 g to 127.5 mg/100 g, and AN15,07,53 from 80.1 mg/100 g to 91.3 mg/100 g). Other studies [43,44] reported similar results, indicating that the increased content of polyphenols is induced in response to the application of water stress. For all the genotypes, the TAC resulted in stability in all the treatments without relevant differences. Fruits of AN15,07,53 from plants under WS1 showed a small decrease in the total antioxidant capacity (237 mg TROLOX/100 g) compared to WW (252.1 mg TROLOX/100 g). On the contrary, fruits of 'Francesca' and 'Lauretta' showed the best performance at WS1 conditions, with, respectively, 322.5 mg TROLOX/100 g and 365.4 mg TROLOX/100 g. Concerning the anthocyanin content, Pel-3-gluc exhibited a small increase in response to stress conditions in all the genotypes (e.g., AN15,07,53 from 18 mg/100 g at WW to 19.3 mg/100 g at WS2 conditions). The overall trend showed a positive outcome for preserving anthocyanin content under WS1 conditions compared to WW because it allows water to be conserved while maintaining a significant level of anthocyanins in the fruit (e.g., for Pel-3-gluc, AN15,07,53 from 18 mg/100 g at WW to 17.9 in WS1 conditions, 'Francesca' from 16 mg/100 g to 17.9 mg/100 g, and 'Lauretta' from 21.5 mg/100 g to 22.6 mg/100 g), as similarly found in other studies [45]. The relation between the fruit Asc.a. content and the treatment showed a small rise among the WW and WS1 in all genotypes. AN15,07,53 reached 20.1 mg/100 g at WW and 22.17 mg/100 g at WS1; Lauretta' reached 30.19 mg/100 g at WW and 32.67 mg/100 g at WS1; and 'Francesca' reached 32.38 mg/100 g at WW and 32.76 mg/100 g at WS1. The Fol. content showed small differences only in 'Francesca' fruit (from 38.4 mg/100 g at WW and 40.6 mg/100 g at WS1) and AN15,07,53 (28.6 mg/100 g at WW and 27.2 mg/100 g at WS1), highlighting an overall trend to uphold the folic acid derivates independently via the given treatment. Chl.A., Caff.A., and Ell.A. did not show differences and remained constant between treatments without statistical significance. The only exception was Ell.A. in 'Francesca' under WS2 conditions, with 10.4 mg/100 g, in contrast to WW and WS1. The response to drought conditions involves several changes in plant physiology. Nonetheless, WS1 demonstrated an adequate preservation of the molecules without compromising other crucial parameters, like the fruits' production. Furthermore, the cultivars stand with higher quality in comparison to the selection, and 'Lauretta' displayed a higher efficiency in terms of water use compared to the others. Finally, we set up a Pearson correlational analysis (Figure 2), providing the positive or negative relationship among all the antioxidant compounds. The map takes into consideration the relations among the molecules analyzed in all possible conditions (WW, WS1, and WS2) to create a coherent trend among the molecule's concentration. The correlational analysis provided a brief, remarkable positive relationship among different molecules. In fact, narrow correlations were found among TAC-TPH and anthocyanins–FOL due to the active mechanism of free radicals scavenging [46]. As expected, the phenolic acids likewise were strictly related. On the other hand, the opposite trend was evident in the relation between Caff.A. and both Asc.A. and Pel.3-gluc and Pel.3-rut and Chl.A. Interestingly, the positive correlation among TPH and all the other compounds analyzed indicates the fact that the Folin-Ciocalteau assay is an analytical procedure still valid today, offering a reliable representation of the antioxidant profile of our fruit or vegetables [47,48].

Table 5. Three-way analysis of variance (ANOVA) of total phenolics (TPH), total antioxidant capacity (TAC), pelargonidine 3-glucoside (Pel.3-gluc), cyanidine 3-glucoside (Cya.3-gluc), pelargonidine-3-rutinoside (Pel.3-rut), ascorbic acid (Asc.a.), folates (Fol.), chlorogenic acid (Chl.a.), ellagic acid (Ell.a.), and caffeic acid (Caff.a.). The parameter interaction was considered strongly significant with p values < 0.01 (*), slightly significant with p values < 0.05 (**), and non-significant with p values > 0.05 (N.S.).

Parameters	TPH	TAC	Pel.3-gluc	Cya.3-gluc	Pel.3-rut
Genotype	**	**	**	**	**
Year	N.S.	*	**	*	N.S.
Treatment	**	*	*	**	**
Treatment × Year	**	**	N.S.	**	N.S.
Treatment × Genotype	**	*	N.S.	**	N.S.
Year × Genotype	*	**	**	N.S.	**
Treatment × Year × Genotype	**	**	**	**	**
Parameters	**Asc.a.**	**Fol.**	**Chl.ac.**	**Caf.ac.**	**Ell.ac.**
Genotype	**	**	**	**	**
Year	*	*	N.S.	N.S.	*
Treatment	**	*	N.S.	N.S.	*
Treatment × Year	N.S.	**	**	N.S.	N.S.
Treatment × Genotype	*	**	N.S.	**	**
Year × Genotype	**	**	N.S.	N.S.	**
Treatment × Year × Genotype	N.S.	**	**	**	**

Table 6. Two-way ANOVA Effect of different irrigation levels on the two years' (2022–2023) mean values of total phenolics (TPH), total antioxidant capacity (TAC), pelargonidine 3-glucoside (Pel.3-gluc), cyanidine 3-glucoside (Cya.3-gluc), pelargonidine-3-rutinoside (Pel.3-rut), ascorbic acid (Asc.a.), folates (Fol), chlorogenic acid (Chl.a.), ellagic acid (Ell.a.), and caffeic acid (Caff.a.) ± standard error. Values with the same lowercase letter for the same parameter were not statistically different for Fisher's LSD test (p < 0.05).

Treatment	Genotype	TPH (mg GA/100 g)	TAC (mg TROLOX/100 g)	Pel-3-gluc (mg/100 g)	Cya-3-gluc (mg/100 g)	Pel-3-rut (mg/100 g)
WW	AN15,07,53	80.1 ± 1.21 d	252.1 ± 11.57 bc	18 ± 0.89 bcd	0.61 ± 0.01 c	1.08 ± 0.32 b
WS1	AN15,07,53	81.7 ± 0.99 d	237 ± 18.07 c	17.9 ± 0.97 cd	0.75 ± 0.005 c	1.38 ± 0.048 b

Table 6. *Cont.*

Treatment	Genotype	TPH (mg GA/100 g)	TAC (mg TROLOX/100 g)	Pel-3-gluc (mg/100 g)	Cya-3-gluc (mg/100 g)	Pel-3-rut (mg/100 g)
WS2	AN15,07,53	91.3 ± 1.02 cd	243.6 ± 19.49 c	19.3 ± 0.95 bcd	0.73 ± 0.009 c	1.37 ± 0.057 b
WW	Francesca	103.5 ± 0.62 b	297.6 ± 13.81 abc	16 ± 0.52 d	0.86 ± 0.01 b	1.59 ± 0.05 b
WS1	Francesca	122.8 ± 1.24 a	322.5 ± 20.88 abc	17.9 ± 1.23 cd	0.82 ± 0.02 b	1.46 ± 0.27 b
WS2	Francesca	127.5 ± 0.76 a	297.2 ± 14.56 abc	16.7 ± 0.87 d	0.9 ± 0.01 b	1.43 ± 0.08 ab
WW	Lauretta	107.7 ± 0.77 bc	334 ± 30.37 abc	21.5 ± 0.78 abc	0.75 ± 0.03 b	1.97 ± 0.2 ab
WS1	Lauretta	103 ± 1.23 bc	365.4 ± 29.13 a	22.6 ± 1.5 ab	0.84 ± 0.03 b	2.52 ± 0.26 ab
WS2	Lauretta	122 ± 1.33 a	314.5 ± 10.87 abc	25.4 ± 1.36 a	0.93 ± 0.03 a	3.30 ± 0.23 a
Treatment	**Genotype**	**Asc.A. (mg/100 g)**	**Fol. (μg/100 g)**	**Chl.A. (mg/100 g)**	**Caff.A. (mg/100 g)**	**Ell.A. (mg/100 g)**
WW	AN15,07,53	20.1 ± 0.33 d	28.6 ± 1.38 bc	9.4 ± 0.47 b	0.8 ± 0.01 bc	8.5 ± 0.08 b
WS1	AN15,07,53	22.17 ± 0.33 d	27.2 ± 0.82 c	8.7 ± 0.42 b	0.9 ± 0.11 ab	8.3 ± 0.2 b
WS2	AN15,07,53	22.98 ± 0.31 d	29.9 ± 0.83 bc	9.1 ± 0.47 b	0.8 ± 0.02 abc	8.6 ± 0.19 b
WW	Francesca	32.38 ± 0.61 ab	38.4 ± 2.56 ab	16.7 ± 1.23 a	0.9 ± 0.03 abc	9.1 ± 0.16 b
WS1	Francesca	32.76 ± 0.55 ab	40 ± 6.41 a	17 ± 1.76 a	1 ± 0.04 ab	8.8 ± 0.16 b
WS2	Francesca	32.8 ± 0.42 ab	32 ± 1.06 bc	16.1 ± 1.8 a	1 ± 0.05 a	10.4 ± 0.38 a
WW	Lauretta	30.19 ± 0.39 bc	30.1 ± 0.98 bc	9.3 ± 0.59 b	0.7 ± 0.02 cd	8.6 ± 0.18 b
WS1	Lauretta	32.67 ± 0.36 ab	29.4 ± 1.52 bc	7 ± 0.57 b	0.5 ± 0.009 d	8.9 ± 0.27 b
WS2	Lauretta	34.79 ± 0.38 a	29.7 ± 1.6 bc	7.7 ± 0.29 b	0.6 ± 0.02 cd	8.9 ± 0.32 b

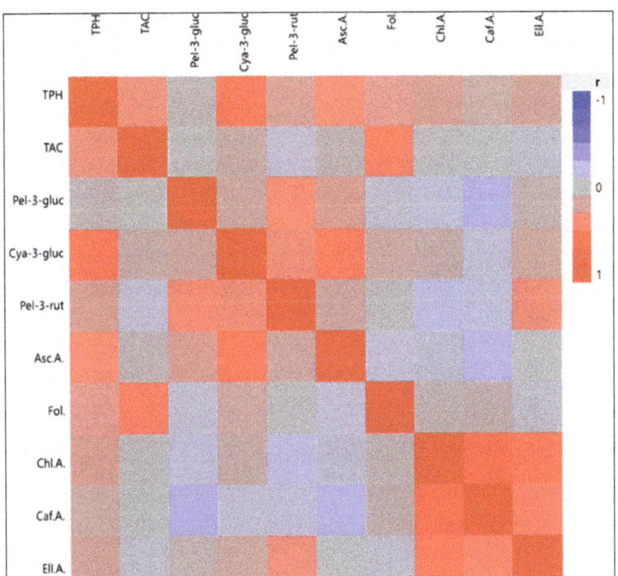

Figure 2. Heatmap of Pearson correlation coefficient of all evaluated traits in the three genotypes at three irrigation levels (WW 100%, WS1 75%, and WS2 56%). The right axis exhibits positive and negative correlations presented in red and blue colors, respectively, to the color scale of total phenolics (TPH), total antioxidant capacity (TAC), pelargonidine 3-glucoside (Pel.3-gluc), cyanidine 3-glucoside (Cya.3-gluc), pelargonidine-3-rutinoside (Pel.3-rut), ascorbic acid (Asc.a.), folates (Fol), chlorogenic acid (Chl.a.), ellagic acid (Ell.a.), and caffeic acid (Caff.a.).

3. Materials and Methods

This two-year experiment was conducted in the experimental greenhouse of the Università Politecnica delle Marche in Ancona (Italy). We tested the following three June bear strawberry genotypes from the UNIVPM breeding program: 'Lauretta', 'Francesca', and AN15,07,53 ('Jonica' × 'Romina'). Generally, in soil farming, Francesca' produced fruits with an excellent nutritional profile, 'Lauretta' exhibited a great compromise among healthy phytochemicals and an appreciable production yield, AN15,07,53 demonstrated great productivity and a set of promising dietary molecules [49]. For two autumn production cultivation cycles, mini-tray plants were planted on 9th September 2022 and 12th September 2023. Both the productions lasted until 3rd November 2022 and 9th November 2023, respectively.

3.1. Experimental Design and Irrigation Scheduling

The experimental irrigation started at the flowering bud's appearance, at stage 55 on the BBCH scale [50], and ended on the last harvest date, at stage 89 on the BBCH scale [50]. For these experiments, a split-plot design was arranged for both years. Specifically, 90 plants were tested (36 plants for genotype). Each block corresponds to a water treatment (WW, WS1, and WS2, described further below), where the genotypes were replicated. The substrate used in these experiments is a mix of peat moss, coconut fiber, silicate, and eco-fiber (a vegetal material derived from renewable material) from a well-known company. The substrate was analyzed (Table 7) to fix the physicochemical and hydraulic properties. Three irrigation treatments were implemented based on the pF values outlined in the table below, representing the logarithmic factor (logarithm base 10) of the matric potential measured in centimeters of water and expressed as the water volume content (wvc), WW (pF1 at 45% wvc), WS1 (pF1-pF1.7 about 30% wvc), and WS2 (with some water shortage points, below pF2, about 21% wvc). Finally, the total water consumption reached 216 L for WW (100%), 160 L for WS1 (75%), and 130 L (60%). The water volume content was determined by adopting a TDR Fieldscout 150 (TDR Fieldscout 150, Plainfield, IL, USA) and a scale for the cross-check. The fertigation program, controlled by a Dosatron® D8R (Dosatron SAS, Tresses, France), was applied for all the treatments. The average temperature and humidity inside the greenhouse were measured using the Testo175H1 sensor (Testo175H1, Lenzkirch, Germany), and both reached, respectively, 24 °C and 73% UR. The physiological status of the plant was measured using the LCi Portable Photosynthesis (LCi Portable Photosynthesis, Hoddesdon, United Kingdom) system, which confirmed the plant's water status in terms of evapotranspiration.

Table 7. Substrate analysis. The values represented show the physical and chemical parameters of the growing media.

Parameter	Unit of Measurement	Standard
Air volume pF1	50.50% v/v	UNI EN 13041:2012
Water volume pF1	45.08% v/v	UNI EN 13041:2012
Air volume pF1,7	71.39% v/v	UNI EN 13041:2012
Water volume pF1,7	24.19% v/v	UNI EN 13041:2012
Air volume pF2	74.15% v/v	UNI EN 13041:2012
Water volume pF2	21.43% v/v	UNI EN 13041:2012
Total Porosity	95.58%	UNI EN 13041:2012
Ph	6.2 pH units	UNI EN 13040:2008 + UNI EN 13037:2012
EC	43 mS/m	UNI EN 13040:2008 + UNI EN 13037:2012
N-NH4	19.96 mg/L	UNI EN 13652:2001
N-NO3	93.71 mg/L	UNI EN 13652:2001
P	38.84 mg/L	UNI EN 13652:2001
Ca	18.12 mg/L	UNI EN 13652:2001
Mg	11.80 mg/L	UNI EN 13652:2001
K	441.47 mg/L	UNI EN 13652:2001
Na	105.45 mg/L	UNI EN 13652:2001

3.2. *Production Evaluation*

The fruits, not rotted, misshaped, or not smaller than Ø < 15 mm, were harvested throughout the fruitification period. All the fruits were weighed to evaluate marketable yield per plant (total marketable yield/plants' number) and average fruit weight (total marketable yield/number of harvested fruits). Finally, water use efficiency (total yield/m^3 water) was determined. The fruits were collected on different days every year, specifically, 7th, 10th, 13th, 18th, and 28th of October and 3rd of November 2022 and 6th, 10th, 13th, 18th, 20th, 25th, and 31st of October 2023. Then, the harvested fruits were divided for organoleptic analyses and nutritional analysis.

3.3. *Strawberry Fruit Quality*

3.3.1. Organoleptic Quality

The fruits were immediately analyzed using the colorimeter CR 400, Konica Minolta, Tokyo, Japan, measuring two points on opposite sides of each fruit using CIELAB values (L*, a*, b*). Then, the samples were frozen at −18 °C until the soluble solids (SS) and titratable acidity analysis using a digital refractometer (PR-101ATAGO, Tokyo, Japan) and the titrating burette, respectively. The TA was calculated as mEQ of NaOH per 100 g of fresh weight (FW) as follows: we added 10 g of distilled water and a few droplets of bromothymol blue (pH indicator) with a 0.1 N NaOH solution of 10 g of strawberry juice as the base [51].

3.3.2. Bioactive Compounds

The strawberry fruits were stored at −18 °C in plastic bags after the harvesting, and 30 undamaged fruits were selected by each harvest day to avoid unwanted oxidative reactions. The frozen fruits were subjected to three different extraction procedures depending on the analytical procedure followed, as described [52]. Shortly, five fruits from the thirty were chosen, chopped, and weighed: 10 g was designated for the methanolic extract suitable for detection of antioxidant capacity (TAC), polyphenols (TPH), anthocyanins, and phenolic acids (Ph.A.); 1 g for detection of vitamin C (Asc.A.); and 2 g for extracting folates (Fol.) (folic acid derivates). Specifically, both TAC and TPH were evaluated through spectrophotometric analyses using a Shimadzu UV-1800 UV/Visible Scanning Spectrophotometer, while Acy, Ph.A., Asc.A., and Fol. were evaluated through liquid chromatography, the HPLC system (Jasco PU-2089 plus), and a Jasco UV-2070 plus ultraviolet (UV) detector (Jasco, Easton, MD, USA). TAC was measured through the 2,20-azino-bis (3-ethylbenzothiazoline-6-sulphonic acid) assay [53,54]. The TPH was calculated using the Folin–Ciocalteu reagent method [55]. The anthocyanins content, specifically for pelargonidin 3-glucoside, pelargonidin 3-rutinoside, and cyanidin 3-rutinoside, was analyzed by following the method outlined by Fredericks et al. [56]. The phenolic acids analyzed were chlorogenic, caffeic acid, and ellagic acid, as described by Schieber et al. [57]. Asc.A. was extracted and analyzed in accordance with the method described [58]. Finally, Fol., as derivates of folic acid, was extracted according to [59].

3.4. *Statistical Analysis*

The results are presented as the average values for the years 2022 and 2023 with the standard error and were subjected to a two-way analysis of variance (ANOVA) at 95% and 99% confidence levels. Significant differences were calculated according to Fisher's LSD test, and differences at $p < 0.05$ were significant. Statistical analyses were performed using Statistica 7 software (StatSoft, TIBCO Software, Palo Alto, CA, USA). The Pearson correlation map was composed with JMP Pro 14.3 software.

4. Conclusions

Improving drought resistance in berry plants is a challenge for plant breeders due to the severity of this abiotic stress. Drought's effects on plants' metabolism have been widely described in many research studies. Nevertheless, little information is currently available on the relationship between the bio-functional compounds of strawberry fruits and the irrigation volumes supplied. The findings provided us with critical insights. Briefly, the water use efficiency highlighted the WS1 treatment as the most convenient irrigation due to the 25% water savings in comparison to full irrigation. At WS1, the WUE increased by 25% for AN15,07,53, 33% for 'Francesca', and 26% for 'Lauretta'. The genotype AN15,07,53 responded in the most efficient way, producing 34 kg of fruits using 1 m^3 of water; followed by 'Lauretta' WS1, with 30 kg of fruits; and 'Francesca', with 18.81 kg of fruits. The WS2 treatment is not optimal due to the critical lack of production in all genotypes, accounting for a lack of 39% in AN15,07,53, 31% in 'Lauretta', and 36% in 'Francesca'. The organoleptic quality of the fruits showed an increase in SS, CA, and color brightness in response to the water deficit. Concerning the bio-functional compounds, TPH proved to be the most affected compound in terms of the water deficit. Specifically, AN15, 07, 53 showed an increase of 2% under WS1 and 12.2% under WS2; 'Francesca' demonstrated a rise of 15.7% under WS1 and 18.8% under WS2; and 'Lauretta' displayed an increase of 3% under WS1 and 15% under WS2. Under normal conditions, 'Lauretta' consistently demonstrated the highest TAC and Acy. Under reduced irrigation (WS1), 'Lauretta' exhibited minimal effects on critical compounds such as Acy and Asc.A. Furthermore, this genotype demonstrated yield stability in response to reduced water availability, suggesting that 'Lauretta' could be an important parent for breeding programs. Ultimately, the sensory and nutritional quality of the fruits is primarily influenced by the genotype [60], which also plays a crucial role in enhancing resilience to environmental stress conditions, including both biotic and abiotic stresses such as water scarcity. Certain molecular mechanisms are essential for improving plant resilience to these stress factors. This study confirms that polyphenol metabolism is directly involved in conferring improved resistance to water stress [33]. Furthermore, other compounds were evidently influenced by the shortening of the water supply, as well as Asc.a. and phenolic acids, even if in a nonsignificant manner. The effective integration of genomic analyses to identify the genes associated with the most influential polyphenols in drought resistance could be vital for optimizing future strawberry breeding programs.

Author Contributions: Conceptualization, D.R. and F.C.; methodology, D.R.; software, D.R.; validation, B.M., F.C. and L.M.; formal analysis, D.R. and R.Q.; investigation, D.R.; resources, F.C.; data curation, D.R.; writing—original draft preparation, D.R.; writing—review and editing, B.M.; visualization, L.M.; supervision, F.C.; project administration, B.M.; funding acquisition, B.M. All authors have read and agreed to the published version of the manuscript.

Funding: This research was funded by Union's Horizon 2020 research and innovation program, grant number 101000747.

Data Availability Statement: The original contributions presented in this study are included in the article. Further inquiries can be directed to the corresponding author.

Conflicts of Interest: The authors declare no conflicts of interest.

References

1. Patra, S.K.; Poddar, R.; Brestic, M.; Acharjee, P.U.; Bhattacharya, P.; Sengupta, S.; Pal, P.; Bam, N.; Biswas, B.; Barek, V.; et al. Prospects of Hydrogels in Agriculture for Enhancing Crop and Water Productivity Under Water Deficit Condition. *Int. J. Polym. Sci.* **2022**, *2022*, 4914836. [CrossRef]
2. Grant, O.M.; Johnson, A.W.; Davies, M.J.; James, C.M.; Simpson, D.W. Physiological and Morphological Diversity of Cultivated Strawberry (*Fragaria* × *ananassa*) in Response to Water Deficit. *Environ. Exp. Bot.* **2010**, *68*, 264–272. [CrossRef]
3. Lavrnić, S.; Zapater-Pereyra, M.; Mancini, M.L. Water Scarcity and Wastewater Reuse Standards in Southern Europe: Focus on Agriculture. *Water Air Soil Pollut.* **2017**, *228*, 251. [CrossRef]
4. Liu, M.; Huang, J.; Sun, A.Y.; Wang, K.; Chen, H. What Roles Can Water-Stressed Vegetation Play in Agricultural Droughts? *Sci. Total Environ.* **2022**, *803*, 149810. [CrossRef]
5. Du, T.; Kang, S.; Zhang, J.; Davies, W.J. Deficit Irrigation and Sustainable Water-Resource Strategies in Agriculture for China's Food Security. *J. Exp. Bot.* **2015**, *66*, 2253–2269. [CrossRef]
6. Anjum, S.A.; Xie, X.Y.; Wang, L.C.; Saleem, M.F.; Man, C.; Lei, W. Morphological, physiological and biochemical responses of plants to drought stress. *Afr. J. Agric. Res.* **2011**, *6*, 2026–2032.
7. Yang, H.; Huntingford, C.; Wiltshire, A.; Sitch, S.; Mercado, L. Compensatory Climate Effects Link Trends in Global Runoff to Rising Atmospheric CO_2 Concentration. *Environ. Res. Lett.* **2019**, *14*, 124075. [CrossRef]
8. Yin, J.; Gentine, P.; Zhou, S.; Sullivan, S.C.; Wang, R.; Zhang, Y.; Guo, S. Large Increase in Global Storm Runoff Extremes Driven by Climate and Anthropogenic Changes. *Nat. Commun.* **2018**, *9*, 4389. [CrossRef]
9. Seleiman, M.F.; Al-Suhaibani, N.; Ali, N.; Akmal, M.; Alotaibi, M.; Refay, Y.; Dindaroglu, T.; Abdul-Wajid, H.H.; Battaglia, M.L. Drought Stress Impacts on Plants and Different Approaches to Alleviate Its Adverse Effects. *Plants* **2021**, *10*, 259. [CrossRef]
10. Marcellini, M.; Raffaelli, D.; Mazzoni, L.; Pergolotti, V.; Balducci, F.; Armas Diaz, Y.; Mezzetti, B.; Capocasa, F. Effects of Different Irrigation Rates on Remontant Strawberry Cultivars Grown in Soil. *Horticult.* **2023**, *9*, 1026. [CrossRef]
11. Surendran, U.; Raja, P.; Jayakumar, M.; Subramoniam, S.R. Use of Efficient Water Saving Techniques for Production of Rice in India under Climate Change Scenario: A Critical Review. *J. Clean. Prod.* **2021**, *309*, 127272. [CrossRef]
12. Lakhiar, I.A.; Yan, H.; Zhang, C.; Wang, G.; He, B.; Hao, B.; Han, Y.; Wang, B.; Bao, R.; Syed, T.N.; et al. A Review of Precision Irrigation Water-Saving Technology Under Changing Climate for Enhancing Water Use Efficiency, Crop Yield, and Environmental Footprints. *Agriculture* **2024**, *14*, 1141. [CrossRef]
13. Molnar, S.; Clapa, D.; Mitre, V. Response of the Five Highbush Blueberry Cultivars to In Vitro Induced Drought Stress by Polyethylene Glycol. *Agronomy* **2022**, *12*, 732. [CrossRef]
14. Hewitt, S.; Hernández-Montes, E.; Dhingra, A.; Keller, M. Impact of Heat Stress, Water Stress, and Their Combined Effects on the Metabolism and Transcriptome of Grape Berries. *Sci. Rep.* **2023**, *13*, 1–14. [CrossRef] [PubMed]
15. FAOSTAT. 2022. Available online: https://www.fao.org/faostat/en/#home (accessed on 18 December 2024).
16. Zahedi, S.M.; Hosseini, M.S.; Fahadi Hoveizeh, N.; Kadkhodaei, S.; Vaculík, M. Physiological. Biochemical Responses of Commercial Strawberry Cultivars under Optimal and Drought Stress Conditions. *Plants* **2023**, *12*, 495. [CrossRef]
17. Renzi, J.P.; Coyne, C.J.; Berger, J.; vonWettberg, E.; Nelson, M.; Ureta, S. How Could the Use of Crop Wild Relatives in Breeding Increase the Adaptation of Crops to Marginal Environments? *Front. Plant Sci.* **2022**, *13*, 886162. [CrossRef]
18. Fan, Z.; Whitaker, V.M. Genomic Signatures of Strawberry Domestication and Breeding. *bioRxiv* **2023**. [CrossRef]
19. Rehman, A.U.; Iso-Touru, T.; Junkers, J.; Rantanen, M.; Karhu, S.; Fischer, D.; Alsheikh, M.; Hjeltnes, S.H.; Mezzetti, B.; Davik, J.; et al. Multi-model GWAS reveals key loci for horticultural traits in reconstructed garden strawberry. *Physiol. Plant.* **2024**, *176*, e14440. [CrossRef]
20. Prohaska, A.; Petit, A.; Lesemann, S.; Rey-Serra, P.; Mazzoni, L.; Masny, A.; Sánchez-Sevilla, J.F.; Potier, A.; Gaston, A.; Klamkowski, K.; et al. Strawberry Phenotypic Plasticity in Flowering Time Is Driven by the Interaction between Genetic Loci and Temperature. *J. Exp. Bot.* **2024**, *75*, erae279. [CrossRef]
21. Sun, S.; Li, Q.; Li, J.; Wang, G.; Zhou, S.; Chai, R.; Hua, W.; Deng, P.; Wang, J.; Lou, W. Revisiting the Evolution of the 2009–2011 Meteorological Drought over Southwest China. *J. Hydrol.* **2019**, *568*, 385–402. [CrossRef]
22. Wang, W.; Hu, W.; Ding, T.; Ye, X.; Liu, D. Shelf-life prediction of strawberry at different temperatures during storage using kinetic analysis and model development. *J. Food Process. Preserv.* **2018**, *42*, e13693. [CrossRef]
23. Marcellini, M.; Mazzoni, L.; Raffaelli, D.; Pergolotti, V.; Balducci, F.; Capocasa, F.; Mezzetti, B. Evaluation of Single-Cropping under Reduced Water Supply in Strawberry Cultivation. *Agronomy* **2022**, *12*, 1396. [CrossRef]
24. Aly, E.M.; Mohamed, M. Sustainable Ecology Strawberry Production Via Different Soilless Culture Techniques Under Urban Conditions. *Egypt. J. Hortic.* **2020**, *47*, 205–218. [CrossRef]
25. Ödemiş, B.; Candemir, D.K.; Evrendilek, F. Responses to Drought Stress Levels of Strawberry Grown in Greenhouse Conditions. *Hortic. Stud.* **2020**, *37*, 113–122. [CrossRef]

26. Gao, F.; Zheng, R.; Guo, J.X.; Li, Y.Z.; Dong, S.K.; Shen, Y.Y.; Li, X.L. Physiological Response of Strawberry to Water Stress under Different Deficit Irrigations. *Grapes Berries J. Crop Prod.* **2020**, *40*, 1–6. [CrossRef]
27. Tsormpatsidis, E.; Ordidge, M.; Henbest, R.G.C.; Wagstaffe, A.; Battey, N.H.; Hadley, P. Harvesting Fruit of Equivalent Chronological Age and Fruit Position Shows Individual Effects of UV Radiation on Aspects of the Strawberry Ripening Process. *Environ. Exp. Bot.* **2011**, *74*, 178–185. [CrossRef]
28. Warner, R.; Wu, B.S.; MacPherson, S.; Lefsrud, M. A Review of Strawberry Photobiology and Fruit Flavonoids in Controlled Environments. *Front. Plant Sci.* **2021**, *12*, 611893. [CrossRef]
29. Perin, E.C.; Messias, R.S.; Galli, V.; Borowski, J.M.; de Souza, E.R.; de Avila, L.O.; Bamberg, A.L.; Rombaldi, C.V. Mineral Content and Antioxidant Compounds in Strawberry Fruit Submitted to Drought Stress. *Food Sci. Technol.* **2019**, *39*, 245–254. [CrossRef]
30. Mancini, M.; Mazzoni, L.; Qaderi, R.; Leoni, E.; Tonanni, V.; Gagliardi, F.; Mezzetti, B. Prediction of Soluble Solids Content by Means of NIR Spectroscopy and Relation with *Botrytis cinerea* Tolerance in Strawberry Cultivars. *Horticulturae* **2023**, *9*, 91. [CrossRef]
31. Bordonaba, J.G.; Terry, L.A. Manipulating the Taste-Related Composition of Strawberry Fruits (*Fragaria* × *ananassa*) from Different Cultivars Using Deficit Irrigation. *Food Chem.* **2010**, *122*, 1020–1026. [CrossRef]
32. Adak, N.; Gubbuk, H.; Tetik, N. Yield, Quality, and Biochemical Properties of Various Strawberry Cultivars under Water Stress. *J. Sci. Food Agric.* **2018**, *98*, 304–311. [CrossRef] [PubMed]
33. Dixon, G.R.; Aldous, D.E. *Horticulture: Plants for People and Places*; Springer: Dordrecht, The Netherlands, 2013; Volume 1. [CrossRef]
34. Wallace, T.C.; Bailey, R.L.; Blumberg, J.B.; Burton-Freeman, B.; Chen, C.Y.O.; Crowe-White, K.M.; Drewnowski, A.; Hooshmand, S.; Johnson, E.; Lewis, R.; et al. Fruits, Vegetables, and Health: A Comprehensive Narrative, Umbrella Review of the Science and Recommendations for Enhanced Public Policy to Improve Intake. *Crit. Rev. Food Sci. Nutr.* **2020**, *60*, 2174–2211. [CrossRef] [PubMed]
35. Krishna, P.; Pandey, G.; Thomas, R.; Parks, S. Improving Blueberry Fruit Nutritional Quality through Physiological and Genetic Interventions: A Review of Current Research and Future Directions. *Antioxidants* **2023**, *12*, 810. [CrossRef] [PubMed]
36. Fernández-Ríos, A.; Laso, J.; Aldaco, R.; Margallo, M. Superfoods: A Super Impact on Health and the Environment? *Curr. Opin. Environ. Sci. Health* **2023**, *31*, 100410. [CrossRef]
37. Qaderi, R.; Mazzoni, L.; Tonanni, V.; Gagliardi, F.; Capriotti, L.; Pergolotti, V.; Diaz, Y.A.; Capocasa, F.; Mezzetti, B. Investigating the Tolerance of Different Strawberry Cultivars to Botrytis Cinerea Infection and Its Relation with Fruit Quality. *J. Berry Res.* **2024**, *14*, 89–103. [CrossRef]
38. Bezerra, M.; Ribeiro, M.; Cosme, F.; Nunes, F.M. Overview of the Distinctive Characteristics of Strawberry, Raspberry, and Blueberry in Berries, Berry Wines, and Berry Spirits. *Comprehen. Rev. Food Sci. Food Saf.* **2020**, *19*, 13354. [CrossRef]
39. Kalt, W.; Cassidy, A.; Howard, L.R.; Krikorian, R.; Stull, A.J.; Tremblay, F.; Zamora-Ros, R. Recent Research on the Health Benefits of Blueberries and Their Anthocyanins. *Adv. Nutr.* **2020**, *11*, 224–236. [CrossRef]
40. Alvarez-Suarez, J.M.; Dekanski, D.; Ristić, S.; Radonjić, N.V.; Petronijević, N.D.; Giampieri, F.; Astolfi, P.; González-Paramás, A.M.; Santos-Buelga, C.; Tulipani, S.; et al. Strawberry Polyphenols Attenuate Ethanol-Induced Gastric Lesions in Rats by Activation of Antioxidant Enzymes and Attenuation of MDA Increase. *PLoS ONE* **2011**, *6*, e25878. [CrossRef]
41. Gonçalves, A.C.; Nunes, A.R.; Falcão, A.; Alves, G.; Silva, L.R. Dietary effects of anthocyanins in human health: A comprehensive review. *Pharmaceuticals* **2021**, *14*, 690. [CrossRef]
42. Nurzyńska-Wierdak, R. Phenolic Compounds from New Natural Sources—Plant Genotype and Ontogenetic Variation. *Molecules* **2023**, *28*, 1731. [CrossRef]
43. Ünal, N.; Okatan, V. Effects of Drought Stress Treatment on Phytochemical Contents of Strawberry Varieties. *Sci. Hortic.* **2023**, *316*, 112013. [CrossRef]
44. Ariza, M.T.; Miranda, L.; Gómez-Mora, J.A.; Medina, J.J.; Lozano, D.; Gavilán, P.; Soria, C.; Martínez-Ferri, E. Yield and Fruit Quality of Strawberry Cultivars under Different Irrigation Regimes. *Agronomy* **2021**, *11*, 261. [CrossRef]
45. Ikeda, T.; Suzuki, N.; Nakayama, M.; Kawakami, Y. The Effects of High Temperature and Water Stress on Fruit Growth and Anthocyanin Content of Pot-Grown Strawberry (*Fragaria* × *ananassa* Duch. cv. 'Sachinoka') Plants. *Environ. Control Biol.* **2011**, *49*, 209–215. [CrossRef]
46. Nga, N.T.T.; An, V.T.; Quang, D.D. Antioxidant Properties of Folic Acid: A Dft Study. *Vietnam J. Sci. Technol.* **2018**, *56*, 39.
47. Saeed, A.; Marwat, M.S.; Shah, A.H.; Naz, R.; Zain-Ul-Abidin, S.; Akbar, S.; Khan, R.; Navid, M.T.; Saeed, A.; Bhatti, M.Z. Assessment of Total Phenolic and Flavonoid Contents of Selected Fruits and Vegetables. *Indian J. Tradit. Knowl.* **2019**, *18*, 686–693.
48. Kocaman, B. Effect of Sorbitol Spraying on Chlorophyl, Total Phenolic and Flavonoid in *Fragaria* × *ananassa*. Duch. cv. Albion Leaves. *Turk. J. Agric.-Food Sci. Technol.* **2024**, *12*, 1352–1358. [CrossRef]

49. George, J.; Edwards, D.; Pun, S.; Williams, D. Evaluation of Antioxidant Capacity (ABTS and CUPRAC) and Total Phenolic Content (Folin-Ciocalteu) Assays of Selected Fruit, Vegetables, and Spices. *Int. J. Food Sci.* **2022**, *2022*, 2581470. [CrossRef]
50. Meier, U. *Growth Stages of Mono-and Dicotyledonous Plants: BBCH Monograph*, 2nd ed.; Federal Biological Research Centre for Agriculture and Forestry: Bonn, Germany, 2001; Volume 49, pp. 66–70.
51. Sadler, G.D.; Murphy, P.A. pH and titratable acidity. *Food Anal.* **2010**, *4*, 219–238. [CrossRef]
52. Mezzetti, B.; Balducci, F.; Capocasa, F.; Zhong, C.F.; Cappelletti, R.; Di Vittori, L.; Mazzoni, L.; Giampieri, F.; Battino, M. Breeding Strawberry for Higher Phytochemicals Content and Claim It: Is It Possible? *Int. J. Fruit Sci.* **2016**, *16*, 194–206. [CrossRef]
53. Miller, N.J.; Rice-Evans, C.; Davis, M.J. A novel method for measuring antioxidant capacity and its application to monitoring the antioxidant status in premature neonates. *Clin. Sci.* **1993**, *83*, 407–412. [CrossRef]
54. Re, R.; Pellegrini, N.; Proteggente, A.; Pannala, A.; Yang, M. Antioxidant activity applying an improved ABTS radical cation decolorization assay. *Free Radic. Biol. Med.* **1999**, *26*, 1231–1237. [CrossRef] [PubMed]
55. Slinkard, K.; Singleton, V.L. Total phenol analysis: Automation and comparison with manual methods. *Am. J. Enol. Vitic.* **1977**, *28*, 49–55. [CrossRef]
56. Fredericks, C.H.; Fanning, K.J.; Gidley, M.J.; Netzel, G.; Zabaras, D.; Herrington, M.; Netzel, M. High-Anthocyanin Strawberries through Cultivar Selection. *J. Sci. Food Agric.* **2013**, *93*, 846–852. [CrossRef] [PubMed]
57. Schieber, A.; Keller, P.; Carle, R. Determination of Phenolic Acids and Flavonoids of Apple and Pear by High-Performance Liquid Chromatography. *J. Chromatogr. A* **2001**, *910*, 265–273. [CrossRef] [PubMed]
58. Helsper, J.P.F.G.; de Vos, C.H.R.; Maas, F.M.; Jonker, H.H.; van den Broeck, H.C.; Jordi, W. Response of Selected Antioxidants and Pigments in Tissues of *Rosa hybrida* and *Fuchsia hybrida* to Supplemental UV-A Exposure. *Physiol. Plant.* **2003**, *117*, 171–187. [CrossRef]
59. Strålsjö, L.; Arkbåge, K.; Witthöft, C.; Jägerstad, M. Evaluation of a radioprotein-binding assay (RPBA) for folate analysis in berries and milk. *Food Chem.* **2002**, *79*, 525–534. [CrossRef]
60. Mazzoni, L.; Di Vittori, L.; Balducci, F.; Forbes-Hernández, T.Y.; Giampieri, F.; Battino, M.; Mezzetti, B.; Capocasa, F. Sensorial and Nutritional Quality of Inter and Intra—Specific Strawberry Genotypes Selected in Resilient Conditions. *Sci. Hortic.* **2020**, *261*, 108945. [CrossRef]

plants

MDPI

Article

Subsoiling Before Wheat Sowing Enhances Grain Yield and Water Use Efficiency of Maize in Dryland Winter Wheat and Summer Maize Double Cropping System Under One-Off Irrigation Practice During the Wheat Season

Yanmin Peng, Kainan Zhao, Jun Zhang, Kaiming Ren, Junhao Zhang, Jinhua Guo, Rongrong Wang, Huishu Xiao, Peipei Jiang, Ninglu Xu, Ming Huang, Jinzhi Wu * and Youjun Li *

School of Agriculture, Henan University of Science and Technology, Luoyang 471023, China; pengym_1998@163.com (Y.P.); zhaokainan6@126.com (K.Z.); bjzhangjun@126.com (J.Z.); renkaiming1848@163.com (K.R.); zh15237517960@163.com (J.Z.); guojinhua0723@163.com (J.G.); wangrr934@163.com (R.W.); xiaohuishu9@126.com (H.X.); jiangpeipei202406@163.com (P.J.); xnl_lulu1129@163.com (N.X.); huangming_2003@126.com (M.H.)

* Correspondence: yywujz@126.com (J.W.); lyj@haust.edu.cn (Y.L.)

Abstract: The winter wheat and summer maize double cropping system is the primary cropping pattern for wheat and maize in dryland areas of China. The management of tillage in this system is typically conducted before wheat sowing. However, few studies have validated and quantified the impact of tillage methods before wheat sowing and irrigation practices during the wheat season on the yield formation and water use efficiency of summer maize. Therefore, this study hypothesized that subsoiling before wheat sowing improves maize yield and WUE by enhancing soil moisture retention and plant development. A three-year field experiment with a two-factor split-plot design was conducted at the junction of the Loess Plateau and the Huang-Huai-Hai Plain in China for validation, from 2019 to 2022. Three tillage methods before wheat sowing (RT: rotary tillage; PT: plowing, SS: subsoiling) were assigned to the main plots, and two irrigation practices during wheat growing season (W0: zero-irrigation; W1: one-off irrigation) were assigned to subplots. We measured the soil moisture, grain yield, dry matter accumulation, nitrogen (N), phosphorus (P), and potassium (K) accumulation, and water use efficiency of summer maize. The results indicated that subsoiling before wheat sowing increased soil water storage at the sowing of summer maize, thereby promoting dry matter and nutrient accumulation. Compared to rotary tillage and plowing, subsoiling before wheat sowing increased grain yield and water use efficiency of maize by an average of 19.5% and 21.8%, respectively. One-off irrigation during the wheat season had negative effects on pre-sowing soil water storage and maize productivity in terms of yield and dry matter accumulation. However, subsoiling before wheat sowing can mitigate these negative effects of one-off irrigation. Correlation analysis and path model results indicated that tillage methods before wheat sowing had a greater impact on soil water storage and maize productivity than irrigation practices during wheat growing season. The most direct factor affecting maize yield was dry matter accumulation, whereas the most direct factor affecting water use efficiency was nutrient accumulation. The technique for order preference by similarity to an ideal solution (TOPSIS) comprehensive evaluation indicated that subsoiling before wheat sowing was superior for achieving high maize yield and water use efficiency under the practice of one-off irrigation during the wheat season. These findings offer practical guidance for optimizing soil water use and maize productivity in drylands.

Keywords: tillage methods; irrigation practices; summer maize; grain yield; water use efficiency; PLSPM; TOPSIS

1. Introduction

Maize (*Zea mays* L.) is one of the most important food, feed, and economic crops globally, playing a crucial role in maintaining global food security and economic development [1–3]. In China, the winter wheat (*Triticum aestivum* L.) and summer maize (hereafter referred to as wheat–maize) double cropping system is the primary cropping pattern for wheat and maize. This system helps to enhance the utilization efficiency of resources such as light and temperature [4]. However, in this system, rotary tillage or no-till planting is generally adopted during the maize season for saving time and costs, which weakens the role of soil tillage in improving soil properties for high maize yields [5]. Research has shown that previous crop tillage has residual effects that can regulate soil properties for subsequent maize crops and increase yield [6,7]. Moreover, the wheat–maize double cropping system is an important pattern in the drylands [6]. In this system, water deficiency is the primary factor limiting crop yield, and the agronomic practice that help to improve water scarcity can strongly increase maize productivity [8,9]. In recent years, the progression of the High-Standard Farmland Construction Program in China and around the world has been swift, ensuring at least one-off irrigation (only irrigated once time during the wheat growth stage) in many dryland regions—arid, semi-arid, and semi-humid drought-prone—that previously lacked irrigation [8]. Previous studies have shown that one-off irrigation during the wheat season can significantly improve wheat yield and water use efficiency [8–10]. However, the limited irrigation (one-off irrigation belongs to this scope) during the wheat season may enhance soil water uptake by in-season crops, which may result in dry soil at the sowing and yield reduction of the subsequent maize [11]. Therefore, it is necessary to understand how one-off irrigation during the wheat season affects soil water storage before the sowing of maize and the nutrient accumulation, yield, and water use efficiency of maize, and seek some effective agronomic strategy to compensate for the water deficiency induced by one-off irrigation during the wheat growing season on maize productivity in the dryland wheat–maize double cropping system.

Optimized tillage methods can improve soil physio-chemical properties, thereby enhancing soil water retention capacity [12,13]. Nevertheless, reasonable tillage practices can harmonize the relationships among water, fertilizer, air, and heat in the soil, promote the root growth of maize, enhance the dry matter and nutrient accumulation and utilization in above-ground parts, and improve maize yield and water use efficiency [12,14,15]. An eight-year experiment in the semi-humid region of China demonstrated that, compared to deep plowing, rotary tillage before wheat sowing could loosen the surface soil and increase the moisture content in the deeper soil layers, thereby significantly enhancing the dry matter accumulation and grain yield of maize by 10.7–11.7% and 11.0–12.3%, respectively, in the wheat–maize double cropping system [16]. However, other studies have indicated that long-term rotary tillage before wheat sowing may cause the plow pan layer to become shallower, which restricts nutrient uptake by maize and is detrimental to maize growth [17,18]. Kuang et al. also demonstrated that deep tillage before wheat sowing improved soil properties, leading to the significant increases in the 100-kernel weight and grain yield of subsequent maize by 2.6–4.2% and 8.3–13.1%, respectively, compared to rotary tillage [6]. Similarly, Latifmanesh et al. found that deep tillage before wheat sowing increased yield by 8.0% and 4.4%, respectively, compared to rotary tillage and no-tillage [7]. Conservation tillage is beneficial for increasing soil water retention capacity, providing more water for crop growth [19,20]. Wang et al. found that, in this area, compared to traditional tillage, conservation tillage increased the bulk density of the surface soil but decreased the bulk density of the subsoil, leading to better water storage during irrigation and rainfall in winter wheat fields [21]. Subsoiling, as a conservation tillage method, has been demonstrated to easily create a structure with alternating loosened and compacted soil layers, optimizing soil moisture, fertility, aeration, and thermal regimes in the tillage layer, allowing roots to grow earlier and extend to deeper soil layers, facilitating the utilization of deep soil water and nutrients by crops [22–24]. Sun et al. indicated that subsoiling before wheat sowing decreased the soil bulk density in the 30–40 cm soil layer at summer maize maturity by 3.9%, and increased the soil porosity, root length density, and grain yield of maize by 5.9%, 30.0%, and 4.7%, respectively [25]. Other field studies showed that subsoiling before wheat sowing could break the plow pan, thus allowing water to penetrate deeper into the soil (below 60 cm) during the summer maize growth period, optimizing water supply for maize growth [6], increasing the areas of green leaf, delaying leaf senescence, and achieving higher dry matter accumulation and yield in maize [26].

Irrigation increases soil moisture and alleviates the adverse effects of drought stress on crop growth and increases crop yields [27,28]. Previous studied have shown that the irrigation practices can not only affecting on crop growth in the current season, but also affect the productivity of the subsequent crop in a rotation cropping system [29–32]. Previous studies have shown that limited irrigation during the wheat season increases soil water consumption, which was unfavorable for summer maize growth and significantly reduced maize yield in the wheat–maize double cropping system [29,33]. Conversely, higher residual soil moisture from full irrigation during the wheat season can enhance summer maize yield [34]. In the study area, one-off irrigation during the wheat season can delay wheat leaf senescence [10], allowing the crop to fully utilize soil water and nutrients in later growth stages, and increased wheat yield by 56%, but it leads to dry soil after harvest [8]. Indeed, under limited water supply conditions during the wheat season, water conservation measures can mitigate the negative effects of limited irrigation on summer maize growth [29]. Wei et al. indicated that the combination of subsoiling and proper irrigation can reduce evapotranspiration, increasing crop yield and water use efficiency by 34.2–84.1% and 19.5–54.3%, respectively [35]. However, few studies have validated and quantified the impact of tillage methods before wheat sowing and irrigation practices during the wheat season on the yield formation and water use efficiency of summer maize.

The technique for order preference by similarity to an ideal solution (TOPSIS) is usually used for comprehensive evaluation of different field management practices [36,37]. In this method, the pathway analysis using Partial Least Squares Path Modeling (PLSPM) is used for quantifying the causal relationships and interaction mechanisms among various variables to enhance scientific understanding of crop production [38]. Then, the TOPSIS method comprehensively evaluates each scheme by measuring the distance between each evaluation scheme and the ideal solution [36,39]. Previous studies have used PLSPM to establish a pathway model including crop rotation, soil chemical properties, and maize growth, showing that crop rotation had a limited impact on maize productivity through soil chemical properties [38,40,41]. Many studies have also used the TOPSIS method for the comprehensive evaluation of field management practices based on a series of different indicators [36,37]. They concluded that the best drought-resistant cultivation model for maize planting in the region is plastic film mulching [36], and the optimal irrigation period is root zone alternate irrigation during the seedling stage when mild water deficiency occurs [37]. However, the comprehensive evaluation of the impact of tillage methods before wheat sowing and irrigation practices during the wheat season on the yield formation and water use efficiency of summer maize using the TOPSIS method is still limited.

Therefore, we hypothesized that subsoiling before winter wheat sowing increases soil water storage before summer maize sowing under the practice of one-off irrigation during the wheat season, thereby enhancing the nutrient accumulation, yield, and water use efficiency of maize. We conducted a three-year split-plot field experiment (three tillage methods—rotary tillage, plowing, and subsoiling—before wheat sowing, and two irrigation practices—zero-irrigation and one-off irrigation—during the wheat growing season) in a typical semi-humid drought-prone region in China. This study aimed to assess how subsoiling before winter wheat sowing influences maize yield and WUE under the practice of one-off irrigation during the wheat season. The comprehensive analysis was performed using PLSPM and TOPSIS methods. If this performs well in this region, it could be extended to similar dryland regions where one-off irrigation is assured.

2. Results

2.1. Soil Water Storage

Figure 1 shows that tillage methods before wheat sowing significantly affected soil water storage at the sowing of maize over the three years. Subsoiling before winter wheat sowing significantly increased soil water storage before summer maize sowing, particularly at depths of 60–140 cm (Figure 1A). Over the three years, the average soil water storage in the 0–200 cm layer under subsoiling was significantly increased by 17.2% and 17.5% compared to rotary tillage and plowing, respectively (Figure 1B). The impact of irrigation practices during the wheat season on soil water storage at the sowing of summer maize varied depending on the tillage methods before wheat sowing. Compared to W0, W1 decreased soil water storage in the 0–200 cm layer by 3.9% and 4.3%, respectively under RT and PT, whereas it increased by 3.6% under subsoiling. The treatment of subsoiling before wheat sowing and one-off irrigation during the wheat season (SSW1) increased maize pre-sowing soil water storage in the 80–140 cm soil layers compared to other treatments, especially when rainfall was severely deficient in June 2021. The soil water storage in the 0–200 cm layers under SSW1 was not significantly different from SSW0 in 2021 and 2022, whereas it was significantly higher than other treatments.

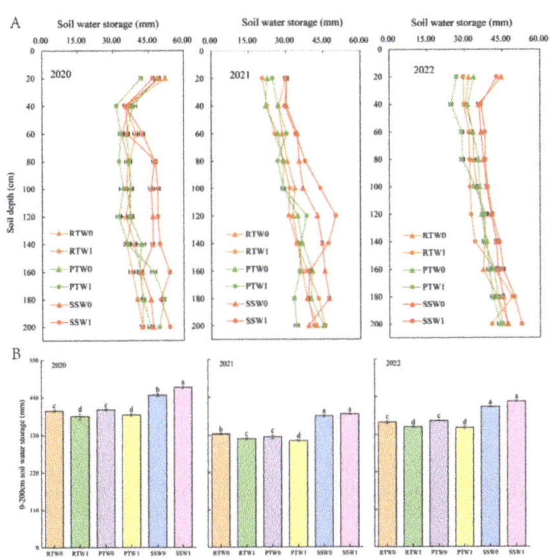

Figure 1. Effects of tillage methods before wheat sowing and irrigation practices during the wheat season on soil water storage in each soil layer (**A**) and total soil water storage in the 0–200 cm layer

(**B**) before maize sowing. RT, PT, and SS represent rotary tillage, plowing, and subsoiling before wheat sowing, respectively. W0 and W1 represent zero-irrigation and one-off irrigation practice during the wheat season, respectively. Different lowercase letters indicate significant differences at $p < 0.05$.

2.2. Above-Ground Dry Matter Accumulation

Figure 2 shows that subsoiling before wheat sowing increased the above-ground dry matter accumulation by 16.3% and 14.7%, respectively, compared to rotary tillage and plowing, in which the dry matter accumulation in stem + leaves and kernels under subsoiling were respectively 15.8% and 6.5% higher than to rotary tillage, and 15.0% and 8.7% higher than plowing. Although W1 had no significant difference in average dry matter accumulation from W0, the effects of irrigation practices on dry matter accumulation varied depending on the tillage method before wheat sowing. Compared to W0, W1 decreased dry matter accumulation under rotary tillage and plowing, while it increased under subsoiling. Particularly in 2020 when rainfall was insufficient, compared to W0, W1 decreased dry matter accumulation by 2.8% for kernels and by 5.2% for stem + leaf under rotary tillage, while they were significantly increased by 5.4% and 4.7% respectively under subsoiling. Overall, SSW1 obtained the highest above-ground dry matter accumulation, which had no significant difference from SSW0 except for 2020. However, over the three years, SSW1 significantly increased the above-ground dry matter accumulation by 14.6–17.6% compared to other treatments.

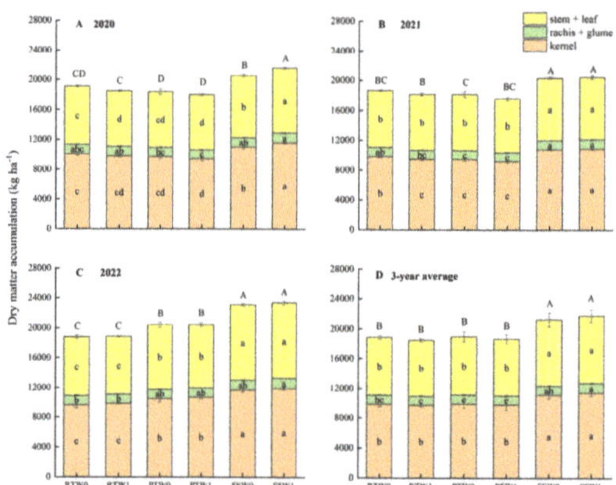

Figure 2. Effects of tillage methods before winter wheat sowing and irrigation practices during the wheat season on the dry matter accumulation of maize in 2020, 2021, 2022 and the three-year average. Note: RT, PT, and SS represent rotary tillage, plowing, and subsoiling before wheat sowing, respectively. W0 and W1 represent zero-irrigation and one-off irrigation during the wheat season, respectively. Different lowercase letters within the same organ indicate significant differences among treatments for organ dry matter accumulation at the $p < 0.05$ level. Different uppercase letters indicate significant differences among treatments for above-ground dry matter accumulation at the $p < 0.05$ level.

2.3. Above-Ground Nutrient Accumulation

Over the three years, subsoiling increased the above-ground nitrogen (N), phosphorus (P), and potassium (K) accumulation by 29.7%, 24.7%, and 23.2%, respectively, compared to rotary tillage, as well as by 25.6%, 22.8%, and 21.3% compared to plowing (Figure 3). On average, irrigation practices during the wheat season did not significantly affect the nutrient accumulation for N, P, and K in maize. However, the effectiveness of irrigation practices varied depending on the tillage method before wheat sowing. Compared to W0, W1 decreased nutrient accumulation under rotary tillage and plowing, whereas it was increased under subsoiling. Particularly, the kernel N, P, and K accumulation respectively decreased by 4.6%,

5.8%, and 5.8% under rotary tillage, and by 3.6%, 7.5%, and 5.5% under plowing, whereas it increased by 3.1%, 7.0%, and 4.6% under subsoiling. The interaction between tillage methods before wheat sowing and irrigation practice during the wheat growing season had a significant effect on above-ground nutrient accumulation in maize at maturity. Specifically, the N, P, and K accumulation under SSW1 were significantly increased by 30.2%, 28.3%, and 28.1%, respectively, compared to other treatments except for SSW0.

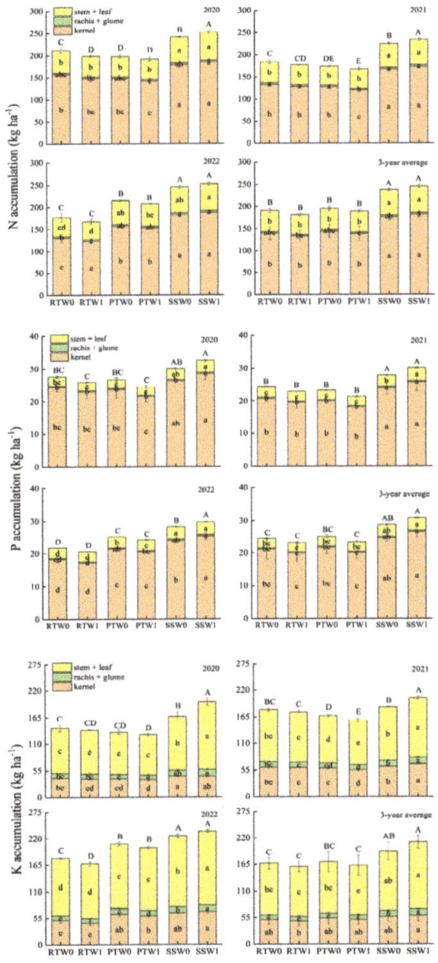

Figure 3. Effects of tillage methods before wheat sowing and irrigation practices during the wheat season on N, P and K accumulation of summer maize. Note: RT, PT, and SS represent rotary tillage, plowing, and subsoiling before wheat sowing, respectively. W0 and W1 represent zero-irrigation and one-off irrigation practice during the wheat season, respectively. Different lowercase letters within the same organ indicate significant differences among treatments for organ N, P and K accumulation at the $p < 0.05$ level. Different uppercase letters indicate significant differences among treatments for above-ground N, P and K accumulation at the $p < 0.05$ level.

2.4. Grain Yield

Both tillage methods before wheat sowing and irrigation practices during the wheat season significantly affected the grain yield and kernels per ear of maize except for 2022

(Table 1). Compared to rotary tillage and plowing, subsoiling increased the kernels per ear by 3.8–25.4% and 4.6–23.2%, the 100-kernel weight by 3.2–10.8% and 7.5–16.4%, and the grain yield by 14.4–29.2% and 13.7–23.3%, respectively. Over the three years, compared to W0, W1 reduced the kernels per ear by 3.0% and 1.6%, 100-kernel weight by 4.2% and 2.6%, respectively under rotary tillage and plowing, while increased by 3.2% and 3.6% under subsoiling. The kernels per ear, 100-kernel weight, and grain yield under SSW1 were significantly higher than other treatments except for SSW0.

Table 1. Effects of tillage methods before wheat sowing and irrigation practices during the wheat season on yield and yield components of maize.

Treatments	Grain Yield (kg ha^{-1})				Kernels Per Ear (kernels ear^{-1})				100-Kernel Weight (g)			
	2020	2021	2022	Average	2020	2021	2022	Average	2020	2021	2022	Average
RTW0	9750 e	9564 b	9474 c	9596 b	454 cd	535 c	514 b	501 b	31 ab	30 bc	33 bc	31 bc
RTW1	9315 f	9239 b	9360 c	9305 b	438 e	526 d	516 b	493 b	31 b	29 bcd	32 c	31 bc
PTW0	9018 c	9024 bc	10,906 d	9649 b	462 c	529 cd	531 b	507 ab	28 c	28 cd	34 bc	30 c
PTW1	8664 d	8554 c	10,501 d	9239 b	446 de	524 d	513 b	494 b	29 c	27 d	33 bc	30 c
SSW0	10,643 b	10,653 a	12,038 a	11,111 a	541 b	544 b	581 a	555 ab	32 ab	31 ab	36 a	33 ab
SSW1	11,160 a	10,944 a	12,297 a	11,467 a	578 a	557 a	589 a	575 a	32 a	33 a	36 a	34 a
Tillage (T)	361.7 **	66.3 **	12.7 **	11.4 **	523.1 **	46.5 **	45.0 **	6.4 *	37.6 **	15.5 **	40.0 **	9.8 **
Irrigation (I)	46.7 **	6.0 *	3.3	0.9	51.5 **	19.0 **	1.3	0.6	0.1	3.6	2.5	1.0
T * I	0.5	0.1	0.3	0.0	4.4 *	1.2	0.7	0.0	2.3	0.4	0.0	1.0

Note: RT, PT, and SS represent rotary tillage, plowing, and subsoiling before wheat sowing, respectively. W0 and W1 represent zero-irrigation and one-off irrigation practice during the wheat season, respectively. Different lowercase letters following the data in the same column indicate significant differences among treatments at the $p < 0.05$ level. * and ** indicate significant variance at the $p < 0.05$ and $p < 0.01$ levels, respectively.

2.5. ET and Water Use Efficiency

Table 2 showed that tillage methods before wheat sowing significantly affected evapotranspiration (ET) during the maize season and water use efficiency (WUE) of maize expect for 2020. On average, compared to rotary tillage, subsoiling did not affect ET, but significantly increased WUE by 20.0%. Compared to plowing, subsoiling significantly reduced ET by 3.0% and increased WUE by 23.6%. Overall, irrigation practices during the wheat season had no significant impact on ET and WUE over the three years. However, compared to W0, W1 tended to decrease WUE under rotary tillage and plowing, whereas increased WUE under subsoiling. Although there was no interaction effect of tillage methods before wheat sowing and irrigation practices during the wheat season on ET, SSW1 increased WUE by 2.0–22.7%, with an average of 18.1%, compared to other treatments.

Table 2. The effects of tillage methods before wheat sowing and irrigation practices during the wheat season on ET during the maize season and water use efficiency of maize.

Treatments	Evapotranspiration (ET, mm)				Water Use Efficiency (kg ha^{-1} mm^{-1})			
	2020	2021	2022	Average	2020	2021	2022	Average
RTW0	454.7 a	433.7 a	471.7 c	453.4 a	21.4 b	22.0 b	20.1 b	21.2 b
RTW1	444.8 a	438.4 a	469.6 c	450.9 a	21 bc	21.1 bc	19.9 b	20.7 bc
PTW0	442.3 a	439.8 a	516.2 a	466.1 a	20.4 c	20.5 cd	21.1 b	20.7 bc
PTW1	443.7 a	440.8 a	498.7 b	461.1 a	19.5 d	19.4 d	21.1 b	20.0 c
SSW0	451.5 a	413.8 b	475.7 c	447.0 a	23.6 a	25.7 a	25.3 a	24.9 a
SSW1	459.1 a	415.7 b	483.4 c	452.7 a	24.3 a	26.3 a	25.4 a	25.4 a
Tillage (T)	2.4	17.5 **	39.5 **	0.3	126.3 **	127.3 **	67.5 **	223.5 **
Irrigation (I)	0	0.4	1.3	0	52.87	2.4	0.0	1.5
T * I	1.3	0.1	4.3 *	0.1	5.3 *	2.8	0.0	3.3

Note: RT, PT, and SS represent rotary tillage, plowing, and subsoiling before wheat sowing, respectively. W0 and W1 represent zero-irrigation and one-off irrigation practice during the wheat season, respectively. Different lowercase letters following the data with the same column indicate significant differences among treatments at the $p < 0.05$ level. * and ** indicate significant variance at the $p < 0.05$ and $p < 0.01$ levels, respectively.

2.6. Correlation and Path Model

2.6.1. Correlation Analysis

Figure 4 showed that there were correlations between every two indicators in term of yield components, plant dry matter accumulation, nutrient accumulation, pre-sowing soil water storage, and yield and water use efficiency in most cases. However, these correlations varied depending on tillage methods before wheat sowing and irrigation practices during the wheat season. The main difference laid in the relationship between soil water storage at sowing and water use efficiency, with the negative correlation under the interactions of tillage methods before wheat sowing and irrigation practices during the wheat season (Figure 4A), and under different tillage methods before wheat sowing (Figure 4B). However, there was a positive correlation between soil water storage at sowing and water use efficiency under irrigation practices during the wheat season (Figure 4C).

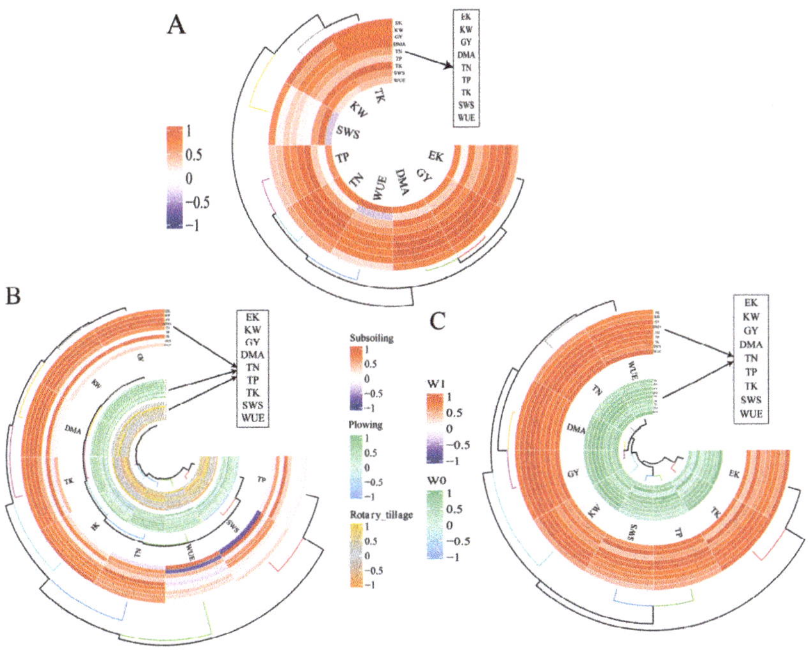

Figure 4. Relationships among the SWS, grain yield, yield components, dry matter accumulation, nutrient accumulation, and water use efficiency under the combination of tillage methods before wheat sowing and irrigation practices during the wheat season (**A**); tillage methods before wheat sowing (**B**) and irrigation practices during the wheat season (**C**).

2.6.2. Partial Least Squares Path Analysis

To further determine the complex relationships among pre-sowing soil water storage, yield traits, dry matter accumulation, nutrient accumulation, and the water use efficiency of maize under different treatments, a partial least squares path model was established (Figure 5A). The goodness-of-fit value was 0.72, indicating a good model fit. These results showed that both tillage methods before wheat sowing and irrigation practice during maize season significantly affected soil water storage at the sowing of maize; greater effectiveness was obtained under tillage methods before wheat sowing than under irrigation practice during maize season. Soil water storage at the sowing of maize directly positively affected yield components and nutrient accumulation. Yield components and nutrient accumulation also had significant positive effects on yield and water use efficiency. Dry matter accumulation had a positive effect on yield but negative effect on water use efficiency. Overall, the model indicated that maize yield was primarily influenced by dry matter accumulation (Figure 5B), and water use efficiency was mainly influenced by nutrient accumulation (Figure 5C).

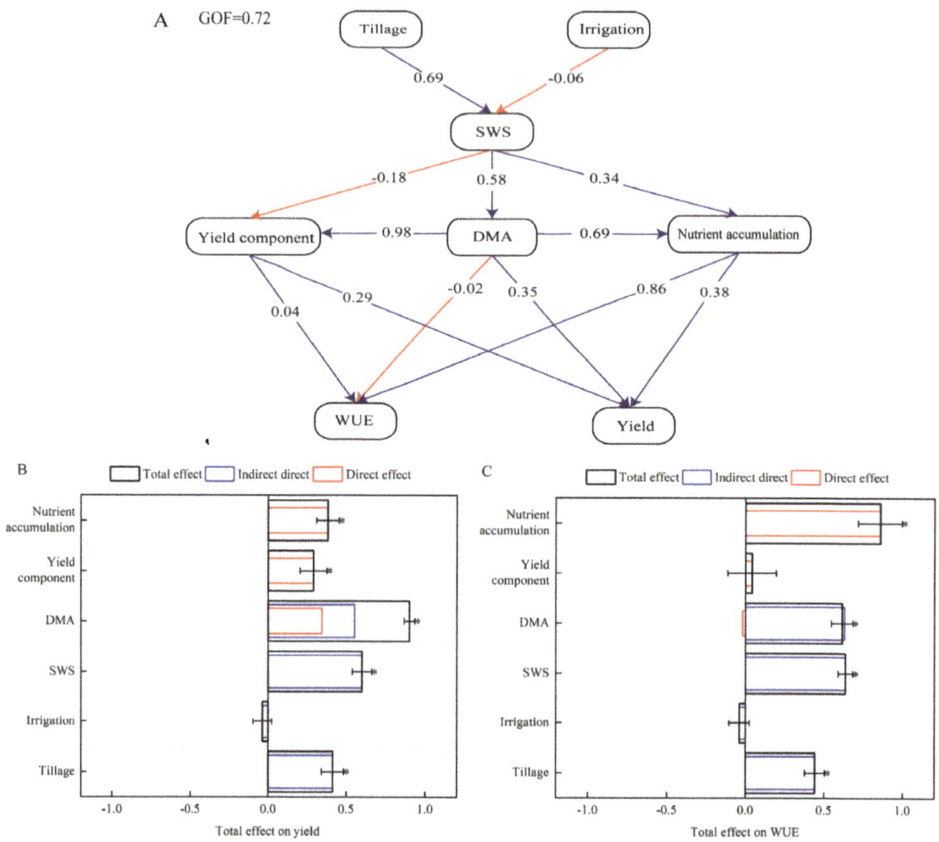

Figure 5. Partial least squares path model (**A**), standardized total effects on grain yield (**B**), and water use efficiency (**C**) based on the model. Note: Path coefficients were calculated after 1000 bootstrap repetitions and are reflected by the width of the arrows, with blue and red indicating positive and negative effects, respectively. The model was evaluated using the goodness-of-fit statistic, and its value was 0.72. * Indicates significance at $p < 0.05$.

2.7. Comprehensive Evaluation

2.7.1. Determination of Weight Using the Entropy Method

The weights of entropy method for yield, kernels per ear, 100-kernel weight, dry matter and N, P, and K accumulation, soil water storage at sowing, and water use efficiency of maize were positive, indicating that the aforementioned indexes were positive indicators. Their weights for 2020 were 0.192, 0.090, 0.090, 0.131, 0.107, 0.066, 0.124, 0.103, and 0.098; for 2021, they were 0.131, 0.064, 0.099, 0.095, 0.129, 0.138, 0.080, 0.156, and 0.108; and for 2022, they were 0.153, 0.080, 0.081, 0.083, 0.070, 0.084, 0.059, 0.273, and 0.116, respectively.

2.7.2. TOPSIS Comprehensive Evaluation

Table 3 indicates that the comprehensive evaluation value of RTW1 was significantly lower than that of RTW0, and that of PTW1 was also significantly lower than that of PTW0. However, the comprehensive evaluation value of SSW1 was significantly higher than that of SSW0. Under SSW1, the comprehensive values of each indicator were closer to the positive ideal solution and farther from the negative ideal solution. The three-year comprehensive evaluation value of SSW1 was significantly higher than other treatments by 12.8–342.9%. Especially in 2020, when rainfall was insufficient, the comprehensive evaluation value of SSW1 was the highest at 0.93. These results suggested that subsoiling before wheat sowing under one-off irrigation during the wheat season better aligns with the goals of high yield and high efficiency of maize in the wheat–maize double cropping system.

Table 3. The degree of fit and ranking under different treatments for summer maize by TOPSIS method.

Treatments	2020				2021				2022			
	di^+	di^-	di	Ranking	di^+	di^-	di	Ranking	di^+	di^-	di	Ranking
RTW0	0.25	0.12	0.33 c	3	0.23	0.12	0.34 c	3	0.31	0.08	0.21 d	5
RTW1	0.29	0.09	0.23 d	4	0.28	0.08	0.22 d	4	0.35	0.04	0.11 e	6
PTW0	0.28	0.07	0.21 d	5	0.28	0.07	0.21 d	5	0.26	0.14	0.34 c	3
PTW1	0.31	0.04	0.12 e	6	0.32	0.03	0.08 e	6	0.33	0.09	0.22 d	4
SSW0	0.11	0.25	0.7 b	2	0.11	0.26	0.71 b	2	0.09	0.31	0.78 b	2
SSW1	0.03	0.33	0.93 a	1	0.05	0.31	0.86 a	1	0.05	0.36	0.88 a	1

Note: di^+: The distance of each evaluation scheme to the positive ideal solution; di^-: The distance of each evaluation scheme to the negative ideal solution; Ci: Closeness coefficient. Different lowercase letters following the data in the same column indicate significant differences among treatments at the $p < 0.05$ level.

3. Discussion

3.1. Soil Water Storage Affected by Subsoiling Before Wheat Sowing Combined with One-Off Irrigation During the Wheat Season

In the double cropping systems, tillage managements for the previous crop have residual effects on soil environment of the subsequent crop [6,7,26]. Previous studies have shown that subsoiling before wheat sowing can reduce soil bulk density and soil penetration resistance during maize growth [42], increase soil porosity, and promote the infiltration of irrigation water and rainfall into the soil [6,7]. Our study also indicated that, compared to rotary tillage and plowing, subsoiling before wheat sowing significantly increased the total water storage in the 0–200 cm soil layer by 17.2% and 17.5%, respectively. The reason for this difference was that subsoiling increased water storage in the 60–140 cm soil layers (Figure 2). This was mainly because subsoiling before winter wheat sowing promotes wheat root growth [43,44], thus the subsoiling-induced dense root system enhances the connectivity among soil pores, guiding and promoting deeper soil water infiltration [6,45]. In addition, subsoiling increases soil stability [46], thus decreasing the evaporation rate of soil moisture from deeper layers [47].

Many studies reported that soil water storage before sowing of the subsequent crop was affected by the irrigation amount and water consumption during the previous crop season [35,48,49]. Some studies showed that the limited irrigation increased water consumption during the wheat season, leading to dry soil before maize sowing [11,33,34]. However, Fang et al. found that increasing irrigation amount during the wheat season increased soil water storage before the sowing of subsequent maize [29]. In our trial, one-off irrigation during the wheat season resulted in the soil storage at the sowing of maize being 5.17 mm lower on average than zero-irrigation over three years. However, the soil water storage change induced by one-off irrigation varied depending on tillage methods before wheat sowing. It decreased by 3.9% and 4.3% under rotary tillage and plowing, respectively, but increased by 3.6% under subsoiling. This was mainly because subsoiling increased soil water holding capacity, enhancing soil water storage during the wheat season [50,51]. Additionally, the results of this trial indicated that tillage methods before wheat sowing had a stronger impact on soil water storage at the sowing of maize than irrigation practices during the wheat season (Figure 6). This may be due to the well-done water protection ability of subsoiling and the limited irrigation amount in our trials (<45 mm). Zhang et al. also indicated that soil storage reduction due to water-saving irrigation can be partially offset by subsoiling [52]. In conclusion, although the coupling effect of tillage methods before wheat sowing and irrigation practices during the wheat season was only significant in some years, the combination of subsoiling before wheat sowing and one-off irrigation during the wheat season increased soil water storage at the sowing of maize, particularly in the 80–140 cm soil layers when rainfall was insufficient (Figure 2).

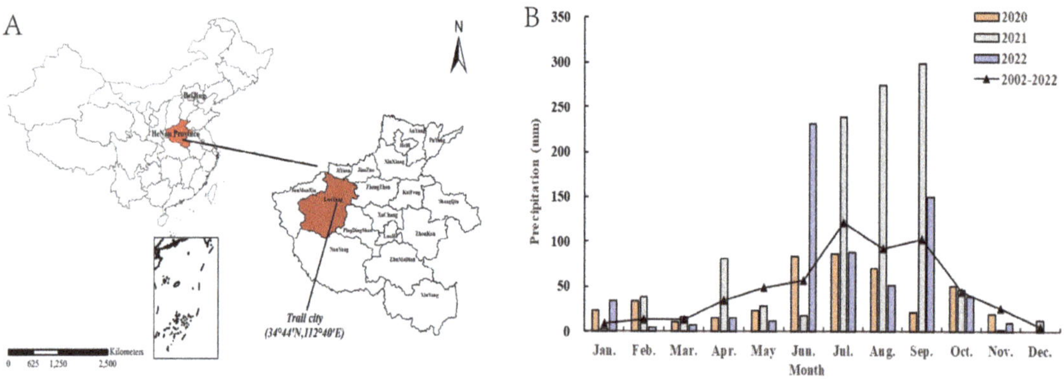

Figure 6. Location (**A**) and precipitation (**B**) of the experimental field in 2020–2022.

3.2. Dry Matter and Nutrient Accumulation Affected by Subsoiling Before Wheat Sowing Combined with One-Off Irrigation During the Wheat Season

Tillage methods alter soil compaction and other properties, thus in turn affecting soil moisture, crop growth and the accumulation of dry matter and nutrients [25,53,54]. As our study showed, compared to rotary tillage and plowing, subsoiling before wheat sowing significantly increased dry matter and N, P, and K accumulation in maize in the double cropping system. This was because subsoiling before wheat sowing can help to break the plow pan in fields [6], promote vertical root distribution in maize, and form a stronger root system, thereby facilitating water and nutrient absorption [55,56], enhancing photosynthesis, and leaf antioxidant characteristics during the grain filling period, ultimately achieving higher dry matter and nutrient accumulation [57,58]. Additionally, our study also indicated a significant correlation between soil moisture before maize sowing and dry matter accumulation. This was mainly ascribed to the soil moisture and other properties improvement induced by subsoiling not only promoting maize root growth but also improving nutrient uptake and accumulation by plants. Particularly, the deep soil moisture improvement under subsoiling plays a critical role in nutrient uptake during the grain filling period, thereby enhancing maize dry matter accumulation [59].

Previous studies have shown that water stress during the maize growth stage disrupted the photosynthetic membrane, reduced chlorophyll content in leaves, and lowered dry matter and nutrient accumulation [60,61]. In this study, compared to zero-irrigation, one-off irrigation tended to decrease the dry matter and nutrient accumulation of maize. Increased water storage through optimizing agronomic measures can mitigate the negative impact of soil water deficiency on the dry matter and nutrient accumulation of maize [29]. Our study showed that one-off irrigation decreased the dry matter and nutrient accumulation under rotary tillage and plowing, but increased the dry matter, N, P, and K accumulation by 5.4%, 3.1%, 7.0%, and 4.6%, respectively, under subsoiling compared to zero-irrigation. These results indicated that subsoiling can mitigate the negative impact of one-off irrigation during the wheat season on maize dry matter and nutrient accumulation.

3.3. Yield and Water Use Efficiency of Maize Affected by Subsoiling Before Wheat Sowing Combined with One-Off Irrigation During the Wheat Season

Kernels per ear and 100-kernel weight directly affect maize yield [62,63]. This study indicated that subsoiling during the wheat season significantly increased kernels per ear and 100-kernel weight compared to rotary tillage and plowing, leading to an increase in yield by 19.5% and 19.5%, respectively. Additionally, this study also found that, one-off irrigation during the wheat season had a significant impact on maize yield and kernels per ear except for 2022, however, it had no significant effect on 100-kernel weight over the three years. This was mainly because irrigation during the wheat season significantly affected soil moisture at the sowing of maize [29,33], and this soil water change can not only regulate the nutrient uptake and photosynthetic rate during the early growth stages [19,64], but also regulate the extension of the pollen tubes, pollination rate, and kernels per ear [65,66], thus affecting the crop growth, and yield formation of maize.

Crop water use efficiency (WUE) is determined by water consumption during the growing period (evapotranspiration) and yield at maturity [62,67]. In our trial, rotary tillage, plowing, and subsoiling before wheat sowing had no significant impact on water consumption during the summer maize growth period. Kuang et al. also showed that rotary tillage and subsoiling during the wheat season had no significant impact on soil water consumption during the maize season [6]. Therefore, in our trails, WUE is only related to yield, and the subsoiling before wheat sowing significantly improved soil water storage before summer maize sowing, which increased maize yield, resulting in a 20.0% and 23.6% increase in WUE compared to rotary tillage and plowing, respectively. The findings confirmed our hypothesis that subsoiling enhanced WUE. However, this result was inconsistent with the field experiment for winter wheat–summer maize system in the North China Plain, where subsoiling increased soil water consumption during the maize season [61].

3.4. Pathway Analysis Using PLSPM and Comprehensive Evaluation

Many studies have shown a significant synergistic effect between soil moisture and maize production [6,68]. However, our correlation analysis showed that the correlation was negative between soil water storage at sowing and water use efficiency of maize under the interaction of the combination of tillage methods before wheat sowing and irrigation practices during the wheat season. This result aligns with correlation analyses under different methods before wheat sowing. The reason maybe that higher soil water storage at sowing leads to shallower maize roots, which is unfavorable for yield formation, thus reducing water use efficiency, especially under drought conditions [69,70].

Previous studies used PLSPM to establish a pathway model including crop rotation, soil chemical properties, and maize growth, showing that crop rotation had a limited impact on maize productivity through soil chemical properties [38,40,41]. The results of our study indicated that tillage methods before wheat sowing and irrigation practices during the wheat season affected maize yield by influencing soil water storage at the sowing of maize. The soil water storage at the sowing of maize had a higher effect on dry matter and nutrient accumulation than on yield components. Therefore, despite the soil water storage at the sowing of maize was negatively correlated with yield components, it was ultimately positively correlated with yield and water use efficiency. The impact of tillage methods before wheat sowing and irrigation practices during the wheat season on yield was mainly through the effect of soil water storage on dry matter accumulation, while the impact on water use efficiency was primarily through nutrient accumulation. Overall, the effect of tillage methods before wheat sowing was much greater than that of irrigation practices during the wheat season, explaining the different impacts of one-off irrigation on maize production under different tillage methods.

TOPSIS has the issue of incomparability among multiple indicators due to different dimensions [39]. Researchers tend to combine the entropy weight method with TOPSIS for comprehensive evaluation of multiple objectives. Previous studies used the entropy weight method and TOPSIS to comprehensively evaluate the drought resistance measures for increasing maize yield [36] and the effects of alternating root-zone irrigation on maize productivity at different growth stages [37]. Our study also used a comprehensive evaluation model combining the entropy weight method and TOPSIS to evaluate the effects of tillage methods before wheat sowing and irrigation practices during the wheat season on soil water storage at sowing, productivity, and water use efficiency of maize. The results indicated that the comprehensive value of SSW1 was the highest, further confirming that subsoiling before wheat sowing is the optimal field management practice for achieving high-yield and high-efficiency production of maize under the one-off irrigation practice during the wheat season in the wheat–maize double cropping season. In addition, conducting quantitative analysis and comprehensive evaluation of the relationship between soil moisture and crop growth under the given experimental conditions inevitably limits the broad applicability of our findings. Thus, if subsoiling before wheat sowing performs well on maize productivity in this region, it could be extended to similar dryland regions where one-off irrigation during the wheat season is assured.

The economic feasibility is a very important index to evaluate the agronomic strategies with different input. Subsoiling requires deeper soil penetration and powerful machinery, which generally leads to higher energy consumption compared to rotary tillage and plowing [71]. This higher energy input may pose economic and environmental challenges. Although previous studies have shown that subsoiling significantly improves crop economic benefits [50], if the subsoiling operation methods and energy efficiency cannot be continuously optimized, its long-term application may be limited. In addition, Repeated subsoiling may alter soil compaction and nutrient cycling over time. Although subsoiling has shown significant short-term soil improvement effects, its long-term sustainability needs further discussion [72]. Therefore, future studies should focus on how to maintain the economic feasibility and long-term sustainability of subsoiling before wheat sowing under the practice of one-off irrigation during the wheat season. For example, adopting more efficient machinery or renewable energy sources, and optimizing the timing and frequency of subsoiling operations are possible solutions. Additionally, integrating subsoiling with other sustainable agricultural practices, such as conservation tillage methods or biodiversity enhancement measures, should be considered. This study also showed that, the adoption of alternating rotary tillage and subsoiling can improve maize yield and water use efficiency of maize. Increasing the intervals of rotary tillage, such as two years of rotary tillage followed by one year of subsoiling or three years of rotary tillage followed by one year of subsoiling, maybe further enhance its long-term sustainability. However, further experimental validation for the proper intervals is needed for specific regions.

4. Materials and Methods

4.1. Experimental Site Description

The present experiment was conducted at Dugou Village, Yichuan County, Henan Province, China (Figure 6A), from October 2019 to October 2022. The experimental site is located in a typical semi-humid drought-prone area with a temperate continental monsoon climate. The average annual temperature is 13.9 °C, and the average annual sunshine duration is 2311 h. The average annual rainfall is 570.8 mm, with approximately 60% concentrated between June and September. During the maize growing seasons from 2020 to 2022, the total rainfall was 263.0 mm, 831.0 mm, and 521.6 mm, respectively (Figure 6B). The soil was classified as heavy loam with pH of 7.6, field capacity of 24.5%, bulk density of 1.33 $g{\cdot}cm^{-3}$, organic matter content of 12.4 g kg^{-1}, total N content of 1.08 g kg^{-1}, available P content of 12.7 mg kg^{-1}, and available K content of 177.1 mg kg^{-1} in the 0–20 cm soil layer at the beginning of the experiment in 2019.

4.2. Experimental Design and Field Managements

The experiment employed a two–factor split-plot design. The main plots included three tillage methods: rotary tillage (RT, with a depth of 15 cm), plowing (PT, with a depth of 30–35 cm), and subsoiling (SS, with a depth of 35–40 cm) before wheat sowing. The depth of specific tillage was set based on the common practice of large-scale households in the study area. The subplots included two irrigation practices: one-off irrigation (W1) and zero-irrigation (W0) during the wheat season. After plowing and subsoiling, a rotary tiller was used to smooth the soil surface. Rotary tillage and plowing were conducted annually, while subsoiling was conducted biennially (subsoiling in 2019 and 2021, rotary tillage in 2020).

A one-off irrigation approach was selected based on the production practice and the results of our previous studies [8,10]. According to the results of previous studies and our filed experiment [8,10], we set the irrigation threshold: irrigation was replenished to 85% of field capacity when the soil moisture content in the 0–40 cm layer first fell below 60% of the field capacity after regreening of wheat. The irrigation amount was calculated according to the method of Ekrenel. [73]:

$$IA = 10 \times \rho b \times H \times (\beta i - \beta j)$$

where IA is the irrigation amount; ρb is the average bulk density ($g{\cdot}cm^{-3}$); H is the average depth of the planned wetting layer (cm); βi is the target soil moisture content (%); βj is the average soil moisture content before irrigation.

The irrigation amount was controlled by a water meter, and the irrigation date was 25 February 2020, 25 February 2021, and 1 March 2022. The irrigation amounts for RT, PT, and SS were 41.1 mm, 43.2 mm, and 37.8 mm in 2020; 39.7 mm, 42.1 mm, and 38.3 mm in 2021; and 37.2 mm, 40.1 mm, and 39.4 mm in 2022. Before summer maize sowing, N fertilizer of 210 kg N ha^{-1}, P fertilizer of 90 kg P_2O_5 ha^{-1}, and K fertilizer of 60 kg K_2O ha^{-1} were evenly applied by hand. After that, a rotary tiller seeder (2BYF-3, Henan Changge Jinfenggu Agricultural Machinery Co., Ltd., Zhengzhou, China) was used for sowing immediately. The N, P, and K fertilizers were urea (N, 46%), triple superphosphate (P_2O_5, 12%), and potassium sulfate (K_2O, 50%), respectively. The maize variety Zhongke Yu 505 was used. The planting density for summer maize was 6×10^4 plants per hectare, with a row spacing of 60 cm and a plant spacing of 27.5 cm. Maize was sown in early and middle June, and harvested in late September or early October. There was no irrigation during the summer maize growing season. Other management practices, such as weeding and pest control, were conducted according to local practices.

4.3. Measurements and Methods

4.3.1. Soil Water

Soil water content was measured by the oven-drying method. At sowing and harvest of maize in 2020–2022, soil samples were randomly collected from each plot at depth ranging from 0 to 200 cm, with increments of 20 cm, using an auger with a diameter of 4.0 cm. Soil samples from the same depth within each plot were thoroughly mixed. Fresh soil samples weighing 50 g $\pm$ 5 g was placed in aluminum boxes and oven-dried at 105 °C for 24 h to measure soil moisture content.

4.3.2. Grain Yield and Yield Components

At harvest, ears of the sample area with 1.8 m $\times$ 2.2 m were manually collected from the center of each plot. Then 10 representative ears were selected to count the number of rows per ear and the number of kernels per row, calculating the kernels per ear. Following harvest, the ears were air-dried, threshed, and the grain obtained was weighed. To accurately determine grain moisture content and dry weight, 100-kernel samples of the air-dried grain were further oven-dried at 90 °C for 30 min and then at a reduced temperature of 65 °C for a duration of 24 h. Grain yield calculations for each plot were standardized to a uniform moisture content of 12.5%, using the air-dried grain weight and its determined water content to adjust the figures according to the method described by Echarte et al. [74].

4.3.3. Dry Matter and Nutrient Accumulation

At maturity of maize in 2020−2022, 10 maize plants were cut from three different rows in each plot. To minimize bias caused by sample collection, these 10 representative plants were sampled within consistent growth conditions, and 3–4 plants from each row. After cutting off the root, samples were separated into three components in terms of stem + leaf, rachis + glume and grain. Sub-samples were oven dried at 105 °C for 30 min and then at 80 °C to determine the water contents and dry weight in each organ. Then, the oven-dried samples of grain, stem + leaf, and rachis + glume were ground with a ball miller (MM400, RETSCH, Haan, Germany), and then digested with $H_2SO_4-H_2O_2$. According to the method of Huang et al. [75], the N and P concentrations in the digest solution were determined using an Auto Analyzer 3 (AA3, Seal Company, Norderstedt, Germany) and K concentration was measured using a flame spectrophotometer (Flame Photometer 410, Sherwood Company, Bury, England). The nutrient (N, P, and K) accumulation in each organ were calculated as the dry weight (kg ha^{-1}) multiplied by the corresponding nutrient concentration (g kg^{-1}), and the dry matter, and nutrient accumulation in above-ground parts (kg ha^{-1}) were calculated from the summed by each organ.

4.3.4. Water Use Efficiency

Soil water storage (SWS), evapotranspiration, and water use efficiency was calculated according to Li [76]:

$$\text{SWS} = 10 \times \text{D} \times \text{H} \times \text{W}$$

where SWS is the soil water storage in the 0–200 cm soil layer (mm); D is the soil bulk density (g·cm^{-3}); H is the soil layer thickness (cm); W is the soil moisture content (%).

$$\text{ET} = \text{P} + \text{U} - \text{R} - \text{F} - \Delta\text{W}$$

where ET is the evapotranspiration during the growing period; P is the precipitation during the maize growing period (mm); R is the runoff (mm); U is the groundwater recharge (mm); F is the deep percolation (mm); ΔW is the difference in soil water storage between the 0–200 cm soil layer before maize sowing and after harvest (mm). When the groundwater table is deeper than 2.5 m, the U value can be ignored. In this experiment, the groundwater was 5 m deep, so all the values for U, F, and R were 0.

$$WUE = Y/ET$$

where Y is the grain yield (kg ha^{-1}), and ET is the evapotranspiration during the maize growing season (mm).

4.4. Calculation of Comprehensive Evaluation Value

4.4.1. Determination of Indicator Weights Using the Entropy Method

Since the evaluation indicators, such as yield and yield components, dry matter accumulation, nutrient accumulation, and water use efficiency are different and cannot be compared directly, the measured data for each indicator were normalized to eliminate the influence of different dimensions. The entropy weight method was then used for objective weight analysis, referring to the method of Zou et al. [77].

4.4.2. Calculation of Comprehensive Evaluation Values for Each Scheme Using TOPSIS

The TOPSIS method evaluates each scheme by measuring the distance to the ideal solution. The best scheme is the one that is closest to the optimal solution and furthest from the worst solution. The comprehensive evaluation value of each treatment is represented by Ci (0 < Ci < 1), where a value closer to 1 indicates that the scheme is more conducive to high maize yield.

4.5. Statistical Analysis

The data were processed using Microsoft Excel 2016 (Microsoft Corp, Redmond, WA, USA) and SPSS 23.0 (SPSS Inc., Chicago, IL, USA) software, and analyzed using ANOVA and post-hoc Duncan's test ($p < 0.05$). Figures were prepared using with Origin 2022 (Origin Lab, Northampton, MA, USA) software. Additionally, R 4.4.1 (R Core Team, University of Auckland, Auckland, New Zealand) was used for Partial Least Squares Path Modeling [78], with validation via the "plspm" package in R (1000 bootstrap repetitions). In this model, we defined pre-sowing soil water storage (SWS), dry matter accumulation (DMA), yield, and water use efficiency as observed variables. Yield components (including kernels per ear and 100-kernel weight) and nutrient accumulation (including N, P, and K accumulation) were defined as latent variables, while tillage and irrigation were considered dummy variables (Figure 6). The direct effects (path coefficients) indicate the direction and strength of the linear relationships between variables [69].

5. Conclusions

This study demonstrated that subsoiling before wheat sowing significantly increased soil water storage at the sowing of maize compared to rotary tillage and plowing, thereby enhancing maize dry matter and nutrient accumulation in above-ground parts, ultimately increasing yield and water use efficiency of maize by 19.5% and 21.8%, respectively. One-off irrigation during the wheat season decreased dry matter and nutrient accumulation, and grain yield of maize, compared to zero-irrigation. Subsoiling before wheat sowing can mitigated these negative effects. Partial least squares path analysis indicated that the impact of tillage methods before wheat sowing and irrigation practices during the wheat season on yield was primarily through the effect of soil water storage on dry matter accumulation, while their influence on water use efficiency was mainly through nutrient accumulation. Furthermore, the TOPSIS comprehensive evaluation results indicated that subsoiling before wheat sowing is an alternative strategy for achieving high-yield and high-efficiency of maize under the practice of one-off irrigation during the wheat season in the wheat–maize double cropping system, it should be a recommended field management practice for the wheat–maize double cropping system in the region where one-off irrigation is assured during the wheat season. These results confirm our hypothesis that subsoiling before wheat sowing significantly enhances maize yield and WUE under one-off irrigation during the wheat season, future studies should focus on the multi-year effects of subsoiling on long-term sustainability in term of economic feasibility, yield stability, and soil prosperities.

Author Contributions: Y.P.: data curation, software, writing—original draft, writing—review and editing. K.Z.: investigation, data curation. J.Z. (Jun Zhang), K.R., J.Z. (Junhao Zhang), J.G.; R.W., H.X., P.J., N.X., M.H. and Y.L.: conceptualization, formal analysis, writing—review & editing. J.W. and Y.L.: conceptualization; funding acquisition, writing—review & editing. All authors have read and agreed to the published version of the manuscript.

Funding: This study was financially supported by the National Key Research and Development Program of China (under Grant No. 2022YFD2300800) and the Science and Technology Research Project of Henan, China (under Grant No. 222102110087; 232102111009).

Data Availability Statement: This study includes all supporting data, which can be obtained from the corresponding authors upon request.

Acknowledgments: The author would like to thank the reviewers for their valuable comments and suggestions for this work.

Conflicts of Interest: The authors declare that they have no known competing financial interests or personal relationships that could have appeared to influence the work reported in this paper.

References

1. Erenstein, O.; Jaleta, M.; Sonder, K.; Mottaleb, K.; Prasanna, B.M. Global maize production, consumption and trade: Trends and R&D implications. *Food Secur.* **2022**, *14*, 1295–1319. [CrossRef]
2. Gao, H.X.; Zhang, C.C.; van der Werf, W.; Ning, P.; Zhang, Z.; Wan, S.B.; Zhang, F.S. Intercropping modulates the accumulation and translocation of dry matter and nitrogen in maize and peanut. *Field Crops Res.* **2022**, *284*, 108561. [CrossRef]
3. Ciampitti, I.A.; Vyn, T.J. A comprehensive study of plant density consequences on nitrogen uptake dynamics of maize plants from vegetative to reproductive stages. *Field Crops Res.* **2011**, *121*, 2–18. [CrossRef]
4. Ding, J.J.; Wu, J.C.; Ding, D.; Yang, Y.H.; Gao, C.M.; Hu, W. Effects of tillage and straw mulching on the crop productivity and hydrothermal resource utilization in a winter wheat-summer maize rotation system. *Agric. Water Manag.* **2021**, *254*, 106933. [CrossRef]
5. Wang, Q.J.; Lu, C.Y.; Li, H.W.; He, J.; Sarker, K.K.; Rasaily, R.G.; Liang, Z.H.; Qiao, X.D.; Li, H.; Mchugh, A.D.J. The effects of no-tillage with subsoiling on soil properties and maize yield: 12-Year experiment on alkaline soils of Northeast China. *Soil Tillage Res.* **2014**, *137*, 43–49. [CrossRef]
6. Kuang, N.K.; Tan, D.C.; Li, H.J.; Gou, Q.S.; Li, Q.Q.; Han, H.F. Effects of subsoiling before winter wheat on water consumption characteristics and yield of summer maize on the North China Plain. *Agric. Water Manag.* **2020**, *227*, 105786. [CrossRef]
7. Latifmanesh, H.; Zheng, C.Y.; Song, Z.W.; Deng, A.X.; Huang, J.L.; Li, L.; Chen, Z.J.; Zheng, Y.T.; Zhang, B.M.; Zhang, W.J. Integrative impacts of soil tillage on crop yield, N use efficiency and greenhouse gas emission in wheat-corn cropping system. *Int. J. Plant Prod.* **2016**, *10*, 317–333. [CrossRef]
8. Zhao, K.N.; Wang, H.T.; Wu, J.Z.; Liu, A.; Huang, X.L.; Li, G.Q.; Wu, S.W.; Zhang, J.; Zhang, Z.W.; Hou, Y.Q.; et al. One-off irrigation improves water and nitrogen use efficiency and productivity of wheat as mediated by nitrogen rate and tillage in drought-prone areas. *Field Crops Res.* **2023**, *295*, 108898. [CrossRef]

9. Wu, J.Z.; Guan, H.Y.; Wang, Z.M.; Li, Y.J.; Fu, G.Z.; Huang, M.; Li, G.Q. Alternative Furrow Irrigation Combined with Topdressing Nitrogen at Jointing Help Yield Formation and Water Use of Winter Wheat under No-Till Ridge Furrow Planting System in Semi-Humid Drought-Prone Areas of China. *Agronomy* **2023**, *13*, 1390. [CrossRef]
10. Huang, M.; Li, W.N.; Hu, C.; Wu, J.Z.; Wang, H.Z.; Fu, G.Z.; Shaaban, M.; Li, G.Q. Coupled one-off alternate furrow irrigation with nitrogen topdressing at jointing optimizes soil nitrate-N distribution and wheat nitrogen productivity in dryland. *Front. Plant Sci.* **2024**, *15*, 1372385. [CrossRef]
11. Feng, Y.P.; Yang, M.; Shang, M.F.; Jia, H.; Chu, Q.Q.; Chen, F. Improving the annual yield of a wheat and maize system through irrigation at the maize sowing stage. *Irrig. Drain.* **2018**, *67*, 755–761. [CrossRef]
12. Liu, Y.; Gao, M.S.; Wu, W.; Tanveer, S.K.; Wen, X.X.; Liao, Y.C. The effects of conservation tillage practices on the soil water-holding capacity of a non-irrigated apple orchard in the Loess Plateau, China. *Soil Tillage Res.* **2013**, *130*, 7–12. [CrossRef]
13. Jabro, J.; Stevens, W.; Iversen, W.; Sainju, U.; Allen, B.; Dangi, S.; Chen, C. Soil-Water Retention Curves and Pore-Size Distribution in a Clay Loam Under Different Tillage Systems. *Land* **2024**, *13*, 1987. [CrossRef]
14. Zhai, L.C.; Wang, Z.B.; Song, S.J.; Zhang, L.; Zhang, Z.B.; Jia, X.L. Tillage practices affects the grain filling of inferior kernel of summer maize by regulating soil water content and photosynthetic capacity. *Agric. Water Manag.* **2021**, *245*, 106600. [CrossRef]
15. Yue, K.; Li, L.L.; Xie, J.H.; Wang, L.L.; Liu, Y.Q.; Anwar, S. Tillage and nitrogen supply affects maize yield by regulating photosynthetic capacity, hormonal changes and grain filling in the Loess Plateau. *Soil Tillage Res.* **2022**, *218*, 105317. [CrossRef]
16. Liu, Y.Y.; Wang, L.; Qian, X.; Li, Z.X.; Gao, Y.B.; Zhang, H.; Liu, S.T.; Liu, K.C.; Zhao, W.J. Effects of wheat stubble tillage methods on the yield formation of no-tillage summer maize. *J. Maize Sci.* **2023**, *31*, 113–119. [CrossRef]
17. Moreira, W.H.; Tormena, C.s.A.; Karlen, D.L.; da Silva, C.l.P.; Keller, T.; Betioli, E. Seasonal changes in soil physical properties under long-term no-tillage. *Soil Tillage Res.* **2016**, *160*, 53–64. [CrossRef]
18. Iqbal, M.; Anwar-ul-Hassan; Ibrahim, M. Effects of tillage systems and mulch on soil physical quality parameters and maize (*Zea mays* L.) yield in semi-arid Pakistan. *Biol. Agric. Hortic.* **2008**, *25*, 311–325. [CrossRef]
19. Bekele, B.; Habtemariam, T.; Gemi, Y. Evaluation of conservation tillage methods for soil moisture conservation and maize grain yield in low moisture areas of SNNPR, Ethiopia. *Water Conserv. Sci. Eng.* **2022**, *7*, 119–130. [CrossRef]
20. Hemmat, A.; Eskandari, I. Conservation tillage practices for winter wheat–fallow farming in the temperate continental climate of northwestern Iran. *Field Crops Res.* **2004**, *89*, 123–133. [CrossRef]
21. Wang, X.B.; Wu, H.J.; Dai, K.; Zhang, D.C.; Feng, Z.H.; Zhao, Q.S.; Wu, X.P.; Jin, K.; Cai, D.X. Tillage and crop residue effects on rainfed wheat and maize production in northern China. *Field Crops Res.* **2012**, *132*, 106–116. [CrossRef]
22. Yan, Z.X.; Gao, C.; Ren, Y.; Zong, R.; Ma, Y.; Li, Q.Q. Effects of pre-sowing irrigation and straw mulching on the grain yield and water use efficiency of summer maize in the North China Plain. *Agric. Water Manag.* **2017**, *186*, 21–28. [CrossRef]
23. Wang, S.B.; Guo, L.L.; Zhou, P.C.; Wang, X.J.; Shen, Y.; Han, H.F.; Ning, T.Y.; Han, K. Effect of subsoiling depth on soil physical properties and summer maize (*Zea mays* L.) yield. *Plant Soil Environ.* **2019**, *65*, 131–137. [CrossRef]
24. Ning, T.Y.; Liu, Z.; Hu, H.Y.; Li, G.; Kuzyakov, Y. Physical, chemical and biological subsoiling for sustainable agriculture. *Soil Tillage Res.* **2022**, *223*, 105490. [CrossRef]
25. Sun, Q.; Sun, W.; Zhao, Z.X.; Jiang, W.; Zhang, P.Y.; Sun, X.F.; Xue, Q.W. Soil Compaction and Maize Root Distribution under Subsoiling Tillage in a Wheat–Maize Double Cropping System. *Agronomy* **2023**, *13*, 394. [CrossRef]
26. Sun, X.F.; Ding, Z.S.; Wang, X.B.; Hou, H.P.; Zhou, B.Y.; Yue, Y.; Ma, W.; Ge, J.Z.; Wang, Z.M.; Zhao, M. Subsoiling practices change root distribution and increase post-anthesis dry matter accumulation and yield in summer maize. *PLoS ONE* **2017**, *12*, e0174952. [CrossRef] [PubMed]
27. Mueller, N.D.; Gerber, J.S.; Johnston, M.; Ray, D.K.; Ramankutty, N.; Foley, J.A. Closing yield gaps through nutrient and water management. *Nature* **2012**, *490*, 254–257. [CrossRef] [PubMed]
28. Seleiman, M.F.; Al-Suhaibani, N.; Ali, N.; Akmal, M.; Alotaibi, M.; Refay, Y.; Dindaroglu, T.; Abdul-Wajid, H.H.; Battaglia, M.L. Drought stress impacts on plants and different approaches to alleviate its adverse effects. *Plants* **2021**, *10*, 259. [CrossRef]
29. Fang, Q.; Wang, Y.Z.; Uwimpaye, F.; Yan, Z.Z.; Li, L.; Liu, X.W.; Shao, L.W. Pre-sowing soil water conditions and water conservation measures affecting the yield and water productivity of summer maize. *Agric. Water Manag.* **2021**, *245*, 106628. [CrossRef]
30. Sun, Z.C.; Zhang, Y.H.; Zhang, Z.; Gao, Y.M.; Yang, Y.M.; Han, M.K.; Wang, Z.M. Significance of disposable presowing irrigation in wheat in increasing water use efficiency and maintaining high yield under winter wheat-summer maize rotation in the North China Plain. *Agric. Water Manag.* **2019**, *225*, 105766. [CrossRef]
31. Nielsen, D.C.; Vigil, M.F.; Anderson, R.L.; Bowman, R.A.; Benjamin, J.G.; Halvorson, A.D. Cropping system influence on planting water content and yield of winter wheat. *Agron. J.* **2002**, *94*, 962–967. [CrossRef]
32. Fan, X.Y.; Schütze, N. Assessing crop yield and water balance in crop rotation irrigation systems: Exploring sensitivity to soil hydraulic characteristics and initial moisture conditions in the North China Plain. *Agric. Water Manag.* **2024**, *300*, 108897. [CrossRef]

33. Zhang, X.Y.; Pei, D.; Chen, S.Y.; Sun, H.Y.; Yang, Y.H. Performance of double-cropped winter wheat–summer maize under minimum irrigation in the North China Plain. *Agron. J.* **2006**, *98*, 1620–1626. [CrossRef]
34. Han, H.F.; Shen, J.Y.; Zhao, D.D.; Li, Q.Q. Effect of irrigation frequency during the growing season of winter wheat on the water use efficiency of summer maize in a double cropping system. *Maydica* **2012**, *56*, 107–112. [CrossRef]
35. Wei, D.; Luo, N.; Zhu, Y.P.; Wang, P.; Meng, Q.F. Diverse water management in a preceding wheat crop does not affect maize yield but increases inter-annual variability: A six-year field study. *Field Crops Res.* **2023**, *302*, 109039. [CrossRef]
36. Liang, C.; Yu, S.C.; Zhang, H.J.; Wang, Z.Y.; Li, F.Q. Economic evaluation of drought resistance measures for maize seed production based on TOPSIS model and combination weighting optimization. *Water* **2022**, *14*, 3262. [CrossRef]
37. Cheng, M.H.; Wang, H.D.; Fan, J.L.; Zhang, F.C.; Wang, X.K. Effects of soil water deficit at different growth stages on maize growth, yield, and water use efficiency under alternate partial root-zone irrigation. *Water* **2021**, *13*, 148. [CrossRef]
38. Zhang, Z.G.; An, J.A.; Xiong, S.W.; Li, X.F.; Xin, M.H.; Wang, J.; Han, Y.C.; Wang, G.P.; Feng, L.; Lei, Y.P. Orychophragmus violaceus-maize rotation increases maize productivity by improving soil chemical properties and plant nutrient uptake. *Field Crops Res.* **2022**, *279*, 108470. [CrossRef]
39. Gurmani, S.H.; Chen, H.Y.; Bai, Y.H. Dombi operations for linguistic T-spherical fuzzy number: An approach for selection of the best variety of maize. *Soft Comput.* **2022**, *26*, 9083–9100. [CrossRef]
40. Ai, C.; Zhang, S.Q.; Zhang, X.; Guo, D.D.; Zhou, W.; Huang, S.M. Distinct responses of soil bacterial and fungal communities to changes in fertilization regime and crop rotation. *Geoderma* **2018**, *319*, 156–166. [CrossRef]
41. Xu, F.; Liu, Y.L.; Du, W.C.; Li, C.L.; Xu, M.L.; Xie, T.C.; Yin, Y.; Guo, H.Y. Response of soil bacterial communities, antibiotic residuals, and crop yields to organic fertilizer substitution in North China under wheat–maize rotation. *Sci. Total Environ.* **2021**, *785*, 147248. [CrossRef]
42. Ren, B.Z.; Li, X.; Dong, S.T.; Liu, P.; Zhao, B.; Zhang, J.W. Soil physical properties and maize root growth under different tillage systems in the North China Plain. *Crop J.* **2018**, *6*, 669–676. [CrossRef]
43. Izumi, Y.; Yoshida, T.; Iijima, M. Effects of subsoiling to the non-tilled field of wheat-soybean rotation on the root system development, water uptake, and yield. *Plant Prod. Sci.* **2009**, *12*, 327–335. [CrossRef]
44. Olesen, J.r.E.; Munkholm, L.J. Subsoil loosening in a crop rotation for organic farming eliminated plough pan with mixed effects on crop yield. *Soil Tillage Res.* **2007**, *94*, 376–385. [CrossRef]
45. Jiang, X.J.; Liu, W.J.; Chen, C.F.; Liu, J.Q.; Yuan, Z.Q.; Jin, B.C.; Yu, X.Y. Effects of three morphometric features of roots on soil water flow behavior in three sites in China. *Geoderma* **2018**, *320*, 161–171. [CrossRef]
46. Czyz, E.; Dexter, A. Soil physical properties as affected by traditional, reduced and no-tillage for winter wheat. *Int. Agrophysics* **2009**, *23*, 319–326. [CrossRef]
47. Wei, S.Y.; Kuang, N.K.; Jiao, F.L.; Zong, R.; Li, Q. Exploring the effects of subsoiling tillage under various irrigation regimes on the evapotranspiration and crop water productivity of winter wheat using RZWQM2. *Agric. Water Manag.* **2023**, *289*, 108531. [CrossRef]
48. Mahdavimanshadi, M.; Fan, N. Multistage Stochastic Optimization for Semi-arid Farm Crop Rotation and Water Irrigation Scheduling Under Drought Scenarios. *J. Agric. Biol. Environ. Stat.* **2024**, 1–24. [CrossRef]
49. KC, B.; Mohssen, M.; Chau, H.; Curtis, A.; Cuenca, R.; Bright, J.; Srinivasan, M.; Safa, M. Irrigation strategies for rotational grazing pasture in Canterbury, New Zealand, and impacts on irrigation efficiency. *Irrig. Drain.* **2018**, *67*, 779–789. [CrossRef]
50. He, J.; Li, H.W.; Wang, X.Y.; McHugh, A.; Li, W.Y.; Gao, H.W.; Kuhn, N. The adoption of annual subsoiling as conservation tillage in dryland maize and wheat cultivation in northern China. *Soil Tillage Res.* **2007**, *94*, 493–502. [CrossRef]
51. Yang, Y.H.; Li, M.J.; Wu, J.C.; Pan, X.Y.; Gao, C.M.; Tang, D.W. Impact of combining long-term subsoiling and organic fertilizer on soil microbial biomass carbon and nitrogen, soil enzyme activity, and water use of winter wheat. *Front. Plant Sci.* **2022**, *12*, 788651. [CrossRef]
52. Zhang, D.; Li, X.D.; Wang, H.G.; Li, H.R.; Li, R.Q.; Batchelor, W.D.; Ju, H.; Li, Y.M. Tillage practices offset wheat yield reductions under limited irrigation regime in the North China Plain. *Soil Tillage Res.* **2023**, *230*, 105687. [CrossRef]
53. Hunt, P.; Bauer, P.; Matheny, T.; Busscher, W. Crop yield and nitrogen accumulation response to tillage of a coastal plain soil. *Crop Sci.* **2004**, *44*, 1673–1681. [CrossRef]
54. Cai, H.G.; Ma, W.; Zhang, X.Z.; Ping, J.Q.; Yan, X.G.; Liu, J.Z.; Yuan, J.C.; Wang, L.C.; Ren, J. Effect of subsoil tillage depth on nutrient accumulation, root distribution, and grain yield in spring maize. *Crop J.* **2014**, *2*, 297–307. [CrossRef]
55. Guan, D.H.; Al-Kaisi, M.M.; Zhang, Y.S.; Duan, L.S.; Tan, W.M.; Zhang, M.C.; Li, Z.H. Tillage practices affect biomass and grain yield through regulating root growth, root-bleeding sap and nutrients uptake in summer maize. *Field Crops Res.* **2014**, *157*, 89–97. [CrossRef]
56. Xu, D.; Mermoud, A. Modeling the soil water balance based on time-dependent hydraulic conductivity under different tillage practices. *Agric. Water Manag.* **2003**, *63*, 139–151. [CrossRef]

57. Lamptey, S.; Li, L.L.; Xie, J.; Coulter, J.A. Tillage system affects soil water and photosynthesis of plastic-mulched maize on the semiarid Loess Plateau of China. *Soil Tillage Res.* **2022**, *196*, 104479. [CrossRef]
58. Li, H.X.; Cheng, Y.F.; Feng, J.X.; Fu, G.L.; Liu, G.L.; Liu, P.; Ren, H.; Wang, H.Z.; Zhao, B.; Li, G. Effects of tillage methods on photosynthetic performance of different functional leaf groups of summer maize in coastal saline-alkali farmland. *Photosynthetica* **2024**, *62*, 339. [CrossRef] [PubMed]
59. Wu, W.M.; Wang, S.J.; Zhang, L.; Li, J.C.; Song, Y.H.; Peng, C.; Chen, X.; Jing, L.L.; Chen, H.J. Subsoiling improves the photosynthetic characteristics of leaves and water use efficiency of rainfed summer maize in the Southern Huang-Huai-Hai Plain of China. *Agronomy* **2020**, *10*, 465. [CrossRef]
60. Song, L.B.; Jin, J.M. Improving CERES-Maize for simulating maize growth and yield under water stress conditions. *Eur. J. Agron.* **2020**, *117*, 126072. [CrossRef]
61. Wang, B.; van Dam, J.; Yang, X.L.; Ritsema, C.; Du, T.S.; Kang, S.Z. Reducing water productivity gap by optimizing irrigation regime for winter wheat-summer maize system in the North China Plain. *Agric. Water Manag.* **2023**, *280*, 108229. [CrossRef]
62. Zhang, X.Y.; Qin, W.M.; Chen, S.Y.; Shao, L.W.; Sun, H.Y. Responses of yield and WUE of winter wheat to water stress during the past three decades—A case study in the North China Plain. *Agric. Water Manag.* **2017**, *179*, 47–54. [CrossRef]
63. Tari, A.F. The effects of different deficit irrigation strategies on yield, quality, and water-use efficiencies of wheat under semi-arid conditions. *Agric. Water Manag.* **2016**, *167*, 1–10. [CrossRef]
64. Zhou, S.W.; Hu, X.T.; Zhou, Z.J.; Wang, W. Improving water use efficiency of spring maize by adopting limited supplemental irrigation following sufficient pre-sowing irrigation in northwest China. *Water* **2019**, *11*, 802. [CrossRef]
65. Luna, V.S.; Figueroa, M.J.; Baltazar, M.B.; Gomez, L.R.; Townsend, R.; Schoper, J. Maize pollen longevity and distance isolation requirements for effective pollen control. *Crop Sci.* **2001**, *41*, 1551–1557. [CrossRef]
66. Cárcova, J.; Otegui, M.E. Ovary growth and maize kernel set. *Crop Sci.* **2007**, *47*, 1104–1110. [CrossRef]
67. Farooq, M.; Wahid, A.; Kobayashi, N.; Fujita, D.; Basra, S.M. Plant drought stress: Effects, mechanisms and management. *Sustain. Agric.* **2009**, *29*, 185–212. [CrossRef]
68. Li, W.; Qin, T.L.; Liu, S.S.; Yang, Y.H.; Liu, H.X.; Xu, S. Driving factors analysis of soil respiration in China ecosystems. *Plant Soil* **2024**, *502*, 1–21. [CrossRef]
69. Wang, X.B.; Zhou, B.Y.; Sun, X.F.; Yue, Y.; Ma, W.; Zhao, M. Soil tillage management affects maize grain yield by regulating spatial distribution coordination of roots, soil moisture and nitrogen status. *PLoS ONE* **2015**, *10*, e0129231. [CrossRef] [PubMed]
70. Lu, H.D.; Xia, Z.Q.; Fu, Y.F.; Wang, Q.; Xue, J.Q.; Chu, J. Response of soil temperature, moisture, and spring maize (*Zea mays* L.) root/shoot growth to different mulching materials in semi-arid areas of Northwest China. *Agronomy* **2020**, *10*, 453. [CrossRef]
71. Shi, J.L.; Li, S.; Tian, X.H. Integrated crop residue and subsoiling management strategies influence soil quality and agricultural sustainability. *Agron. J.* **2021**, *113*, 537–549. [CrossRef]
72. Yang, Y.H.; Wu, J.C.; Du, Y.L.; Gao, C.M.; Pan, X.Y.; Darrel, W.S. Martine Var der Ploeg. Short-and long-term straw mulching and subsoiling affect soil water, photosynthesis, and water use of wheat and maize. *Front. Agron.* **2021**, *3*, 708075. [CrossRef]
73. Ekren, S.; Sönmez, Ç.; Özçakal, E.; Kurttaş, Y.S.K.; Bayram, E.; Gürgülü, H. The effect of different irrigation water levels on yield and quality characteristics of purple basil (*Ocimum basilicum* L.). *Agric. Water Manag.* **2012**, *109*, 155–161. [CrossRef]
74. Echarte, L. Andrade F H. Harvest index stability of Argentinean maize hybrids released between 1965 and 1993. *Field Crops Res.* **2003**, *82*, 1–12. [CrossRef]
75. Huang, M.; Wang, C.H.; Luo, L.C.; Wang, S.; Hui, X.L.; He, G.; Cao, H.B.; Ma, X.L.; Huang, T.M.; Zhao, Y.; et al. Soil testing at harvest to enhance productivity and reduce nitrate residues in dryland wheat production. *Field Crops Res.* **2017**, *212*, 153–164. [CrossRef]
76. Li, J.P.; Wang, Z.M.; Yao, C.S.; Zhang, Z.; Liu, Y.; Zhang, Y.H. Micro-sprinkling irrigation simultaneously improves grain yield and protein concentration of winter wheat in the North China Plain. *Crop J.* **2021**, *9*, 1397–1407. [CrossRef]
77. Zou, Z.H.; Yi, Y.; Sun, J.N. Entropy method for determination of weight of evaluating indicators in fuzzy synthetic evaluation for water quality assessment. *J. Environ. Sci.* **2006**, *18*, 1020–1023. [CrossRef]
78. Tenenhaus, M.; Vinzi, V.E.; Chatelin, Y.-M.; Lauro, C. PLS path modeling. *Comput. Stat. Data Anal.* **2005**, *48*, 159–205. [CrossRef]

plants

MDPI

Article

Nutrient Uptake of Two Semidomesticated *Jaltomata* Schltdl. Species for Their Cultivation

Ignacio Darío Flores-Sánchez [1,*], Manuel Sandoval-Villa [1] and Ebandro Uscanga-Mortera [2]

1 Postgraduate in Edaphology, Colegio de Postgraduados, Mexico-Texcoco Highway, km 36.5, Montecillo, Texcoco 56264, Mexico; msandoval@colpos.mx

2 Postgraduate in Botany, Colegio de Postgraduados, Mexico-Texcoco Highway, km 36.5, Montecillo, Texcoco 56264, Mexico; euscanga@colpos.mx

* Correspondence: flores.ignacio@colpos.mx; Tel.: +52-551-077-6669

Abstract: The nutrient uptake of a species under cultivated conditions is important for program fertilization. The *Jaltomata* genus has two semidomesticated species, *J. procumbens* and *J. tlaxcala*, used as food and considered with potential for their study in controlled environments. The objective of this research was to determine nutrient uptake curves of these species in a greenhouse and using hydroponics. The research was carried out at the Colegio de Postgraduados, Campus Montecillo, Texcoco, State of Mexico, from August to November 2020. The treatments included the following: two species and three electrical conductivity levels: 1, 2, and 3 dS m^{-1}. Nutrients in leaf and total dry matter (TDM) were determined. Variability between species and phenological stages on the nutrient concentration and accumulation of TDM was observed. For macronutrients, *J. procumbens* concentrated in descending order more P from the vegetative stage (4.21–2.43 g kg^{-1} dry matter), and Mg until fructification (4.92–3.26 g kg^{-1} dry matter), for K it was higher at vegetative (52.29 g kg^{-1} dry matter) and harvesting stages (26.05 g kg^{-1} dry matter), and N (23.92 g kg^{-1} dry matter) at flowering; *J. tlaxcala* concentrated more Ca from fructification (10.10–13.85 g kg^{-1} dry matter). For micronutrients, *J. tlaxcala* concentrated more Fe from the vegetative stage (157.7–207.5 mg kg^{-1} dry matter), B and Zn at 23.3–38.4 and 26.04–28.45 mg kg^{-1} dry matter, respectively, from flowering, and Mn (108.4–232.28 mg kg^{-1} dry matter) from fructification. The main structures of TDM accumulation by vegetative stage in *J. procumbens* were the leaf and root (vegetative and flowering), root and stem (fructification), and reproductive structures and root (harvesting); in *J. tlaxcala*, the main structures were the leaf and root (vegetative), root and leaf (flowering and fructification), and root and reproductive structures (harvesting). Due to this variability, specific fertilization programs are required for each species.

Keywords: native plant resources; absorption curves; protected agriculture; hydroponics

1. Introduction

Plant nutrition is one of the most important aspects of crop production [1], since optimal growth and yield will depend on it [2].

The genus *Jaltomata* Schltdl., family Solanaceae, has two recognized diversity centers, South America and Mexico with 63 and 8 species, respectively. They are herbaceous and shrubby with edible fruits [3]. In Mexico, *J. procumbens* of erect growth and blue dark fruit is found as ruderal and arvense (plant located on the edge or inside of crop fields useful for human being); *J. tlaxcala*, recently recognized as a new species [4], is located in specific regions of the country, has a decumbent growth habit with green fruit in the mature stage, and is found as an arvense. They are considered a semidomesticated species and are linked to traditional food production systems in Mexico, characterized by being on hillsides or superficial soils, and with frequently degraded soils [5]. Around 64% of Mexican agricultural lands present a level of degradation, and 28% of farmers report some problem with low fertility of their soils [6]. The fruits of these species are recollected and consumed fresh or after some preparations [7].

Their study is relevant due to their important nutrient content and nutraceutical compounds [8–10], characteristics that could help address the problem of malnutrition mentioned by the World Health Organization [11], which is linked to the reduced diversity of foods in production systems and by the poor supply of vitamins, micronutrients, and nutraceutical compounds, because 60% of the 90% of calories consumed worldwide by the population are provided by the main basic crops rich in carbohydrates (rice, corn, potato, soybeans, and wheat) [12–15].

Also, these species are considered viable for their future use because these meet the criteria to consider new plant resources for food purposes [16]: edible plants for human being, plants with high nutritional value; plants that do not need intermediate process between growing, harvest, and use; and plants preferentially native to Mexico.

From the evaluation of agronomic traits, both species have the potential to be studied in controlled environments [7]; however, based on the literature review, there is still no information on the demand and nutrient uptake for both species.

The dynamic uptake of nutrients by the crop is important, to generate the absorption curves of nutrients for each phenological stage, which allows the identification of the maximum absorption stages during the growing season and for the design of fertilization programs [17].

These curves are determined by genetic factors, phenological stage, and weather conditions [1]; then, it is necessary to determine the nutrient uptake dynamics of *Jaltomata procumbens* and *J. tlaxcala*, according to their nutrimental demand during the growing season, which allows their management under greenhouse and hydroponic conditions, which are different from those present under their natural environment, considering the hypothesis that each species has a different dynamics uptake of macro and micronutrients in each phenological stage.

That is why the objective of this research was to determine the absorption curves of nutrients for both *Jaltomata* species for the vegetative, flowering, fructification, and harvesting stages in greenhouse and hydroponic conditions.

2. Results

2.1. Analysis of Effects on the Nutrient Concentration and Total Dry Matter

The species (Sps) and electrical conductivity (EC) influenced the concentration of macronutrients and micronutrients in the leaf and the total dry matter (TDM), in at least one of the samplings, except for Mo and SPAD readings for Sps and B for EC ($p \leq 0.05$). In the Sps x EC interaction, in at least one of the samplings, statistical differences were registered in macronutrients, micronutrients, and TDM and SPAD readings, which indicated that the EC effect for each variable was not the same within each species. The highest coefficients of variation were recorded in micronutrients and TDM: TDM (21.1%), B (41.2%), Mo (42.5%), and Zn (23.7%) for the first sampling; Fe (25.8%) and Mo (48.8%) in the second sampling; TDM (29.6%) and Mo (53.0%) third sampling; and TDM (32.5%), Fe (29.7%) and Mo (73.4%) for the fourth sampling.

2.2. Macronutrients

For N, except for the vegetative stage, there were statistical differences. In the flowering stage, an increase in both Sps was registered, greater in *J. procumbens* in 10.7% compared to *J. tlaxcala*. Subsequently, there was a reduction in both Sps, with *J. procumbens* maintaining a higher concentration in the fructification and harvest stages than *J. tlaxcala* (Figure 1). For P, the highest absorption was registered in *J. procumbens*, with statistical differences in the vegetative and fructification stages. The uptake dynamic was the same for both Sps, with the highest values in the first stage, gradually reducing until harvest, with the lowest values. For K, except for the flowering stage, there were statistical differences; in the vegetative stage, *J. procumbens* recorded more K than *J. tlaxcala*; from this stage, the uptake was reduced in both Sps; however, it was higher in 50.2% in *J. procumbens* at harvest. For *J. tlaxcala*, this reduction was less accentuated with 23.9%.

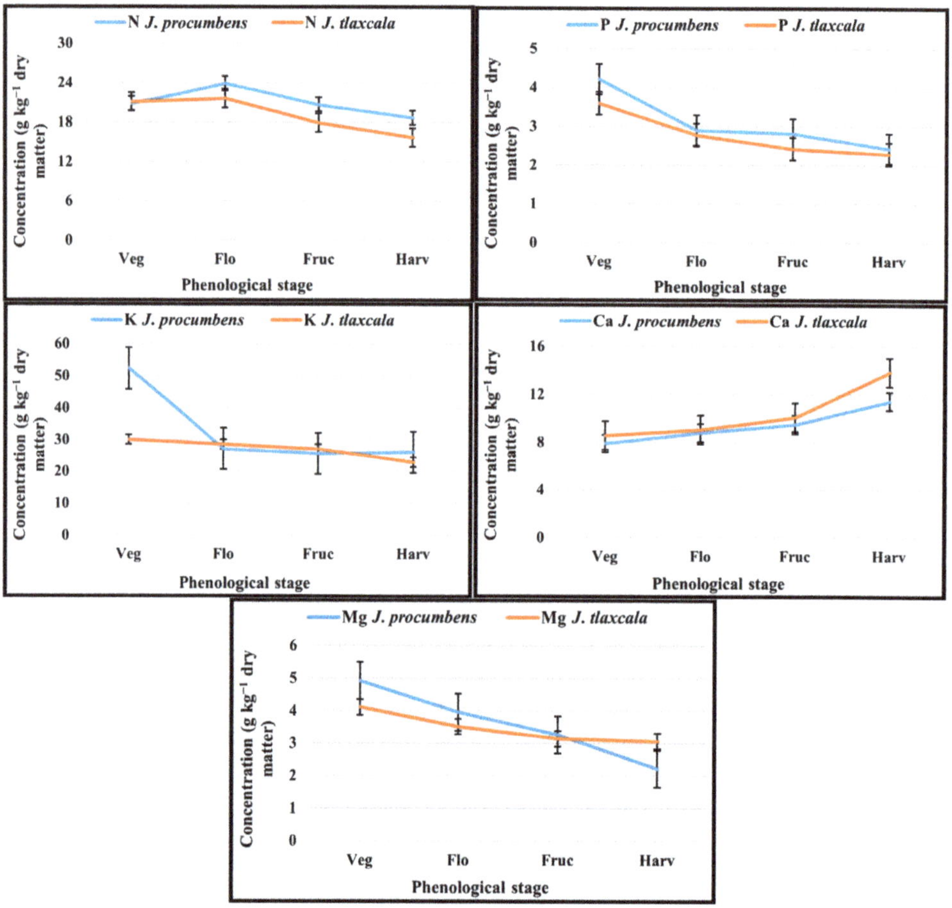

Figure 1. Uptake of N, P, K, Ca, and Mg in *Jaltomata procumbens* and *Jaltomata tlaxcala*. Colegio de Postgraduados, Campus Montecillo, Texcoco, State of Mexico. September 2020–February 2021. The samplings were carried out at 17, 31, 45, and 57 days after transplant (dat), corresponding to the stages: Veg, Vegetative; Flo, Flowering; Fruc, Fructification; and Harv, Harvesting. DMSH: Stage: vegetative N (3.22 g kg^{-1}), P (0.31 g kg^{-1}), K (5.76 g kg^{-1}), Ca (0.68 g kg^{-1}), Mg (0.49 g kg^{-1}); flowering N (1.99 g kg^{-1}), P (0.19 g kg^{-1}), K (2.78 g kg^{-1}), Ca (0.81 g kg^{-1}), Mg (0.40 g kg^{-1}); fructification N (1.51 g kg^{-1}), P (0.24 g kg^{-1}), K (1.14 g kg^{-1}), Ca (0.67 g kg^{-1}), Mg (0.26 g kg^{-1}); harvesting N (1.47 g kg^{-1}), P (0.42 g kg^{-1}), K (3.01 g kg^{-1}), Ca (1.38 g kg^{-1}), Mg (0.43 g kg^{-1}) (Tukey, $p \leq 0.05$).

The Ca uptake increased in each of the phenological stages. The *J. tlaxcala* species statistically exceeded *J. procumbens* at harvest. The highest uptake occurred from fructification, *J. tlaxcala* ended with 21.2% more than *J. procumbens*. For Mg, statistical differences were presented in the vegetative, flowering, and harvest stages; in both Sps, the highest uptake occurred in the first stage, presenting *J. procumbens* the highest value, higher than *J. tlaxcala* by 19.6%; subsequently, a reduction was registered until the last stage, becoming more accentuated in *J. procumbens* with the lowest value compared to *J. tlaxcala*, by 38.5% (Figure 1).

2.3. Micronutrients

For B, there were statistical differences from the flowering stage, with 18.2 mg kg^{-1} for *J. procumbens* and 23.3 mg kg^{-1} for *J. tlaxcala*; subsequently, there was a gradual increase until the harvest stage, where *J. tlaxcala* obtained the highest concentration, higher than *J. procumbens* by 16.6% (Figure 2). For Fe, there were statistical differences in the vegetative stage where *J. tlaxcala* absorbed more Fe than *J. procumbens*, by 30.2%; subsequently, it reduced in both Sps to its lowest point in the flowering and fructification stages; from these stages, there was an increase where *J. tlaxcala* absorbed more Fe than *J. procumbens* by 9.6% for the harvest. The absorption of Mn had a constant increase with statistical differences in the vegetative and harvest stages, registering the highest concentration from fructification at the beginning, with the highest value coming from *J. tlaxcala* in the harvest, by 45.3% more than *J. procumbens*.

For Mo, *J. procumbens* registered the highest amount of this element, with the highest concentration in the flowering stage, higher by 38.0% than *J. tlaxcala*, which registered a decrease respect to the vegetative stage, although there were no statistical differences. For Zn, except for vegetative stage, there were statistical differences in the other stages, where *J. tlaxcala* registered a greater amount of this nutriment in flowering, fructification, and harvest, with respect to *J. procumbens* in 12.7, 12.4, and 18.7%, respectively (Figure 2).

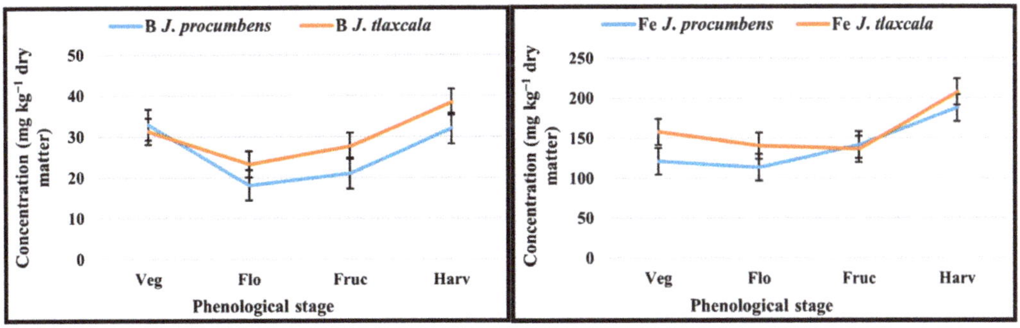

Figure 2. *Cont.*

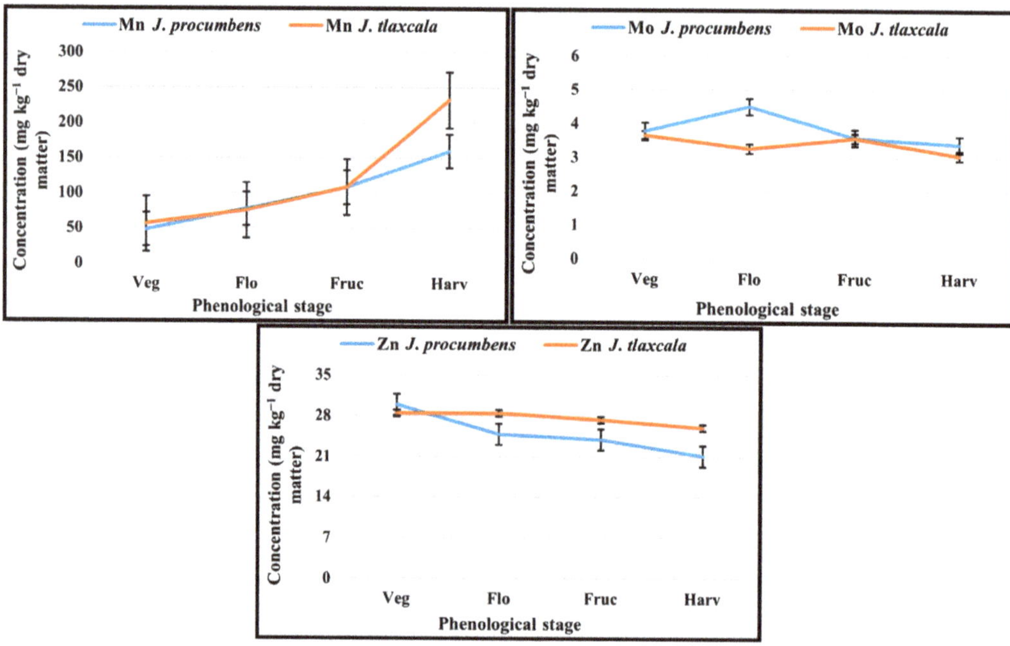

Figure 2. Uptake of B, Fe, Mn, Mo, and Zn in *Jaltomata procumbens* and *Jaltomata tlaxcala*. Colegio de Postgraduados, Campus Montecillo, Texcoco, State of Mexico. September 2020–February 2021. The samplings were carried out at 17, 31, 45, and 57 dat, corresponding to the stages: Veg, Vegetative; Flo, Flowering; Fruc, Fructification; and Harv, Harvesting. DMSH: Stage: vegetative B (13.34 mg kg^{-1}), Fe (22.14 mg kg^{-1}), Mn (9.76 mg kg^{-1}), Mo (1.60 mg kg^{-1}), Zn (7.00 mg kg^{-1}); flowering B (2.19 mg kg^{-1}), Fe (33.11 mg kg^{-1}), Mn (10.47 mg kg^{-1}), Mo (1.91 mg kg^{-1}), Zn (4.22 mg kg^{-1}); fructification B (3.53 mg kg^{-1}), Fe (25.58 mg kg^{-1}), Mn (14.43 mg kg^{-1}), Mo (1.92 mg kg^{-1}), Zn (3.24 mg kg^{-1}); harvest B (6.08 mg kg^{-1}), Fe (59.27 mg kg^{-1}), Mn (36.20 mg kg^{-1}), Mo (2.39 mg kg^{-1}), Zn (4.10 mg kg^{-1}) (Tukey, $p \leq 0.05$).

2.4. Electrical Conductivity

Plant growth with 1 dS m^{-1} registered the highest concentration of Ca in the harvest stage, higher by 23.6% compared to 3 dS m^{-1} EC; for Mg, the highest concentration was recorded in the vegetative stage, with 15.4% more compared to 3 dS m^{-1} EC, indicating that the concentration decreases as EC is increased. For Mn, the highest concentration was registered in the fructification stage, higher by 19.9% compared to 3 dS m^{-1} EC (Table 1).

Table 1. Comparison of means on the concentration of N, P, K, Ca, Mg, Fe, Mn, Mo, and Zn of *Jaltomata procumbens* and *Jaltomata tlaxcala* cultivated at three electrical conductivity levels of nutrient solution. Colegio de Postgraduados, Campus Montecillo, Texcoco, State of Mexico. September 2020–February 2021. Data taken between 17 and 129 days after transplant (dat).

FV	Macronutrients (g kg^{-1} Dry Matter)				Micronutrients (mg kg^{-1} Dry Matter)	
EC (dS m^{-1})	Vegetative Stage (17 dat)				Vegetative Stage (17 dat)	
	N	P	Ca	Mg	Fe	Mo
1	17.8 b	3.5 b	7.4 b	4.8 a	118.3 b	1.9 b
2	21.6 ab	4.1 a	8.7 a	4.7 a	146.6 ab	4.1 a
3	23.2 a	4.1 a	8.6 a	4.0 b	153.4 a	5.2 a

Table 1. *Cont.*

FV	Macronutrients (g kg^{-1} Dry Matter)			Micronutrients (mg kg^{-1} Dry Matter)		
EC (dS m^{-1})	Vegetative Stage (17 dat)			Vegetative Stage (17 dat)		
	Flowering stage (31 dat)			Flowering stage (31 dat)		
	N	P	Mg	Mn	Mo	Zn
1	21.2 b	2.3 c	4.3 a	94.4 a	3.2 b	24.2 b
2	22.6 ab	2.8 b	3.3 b	75.4 b	2.5 b	25.8 ab
3	24.3 a	3.4 a	3.6 b	59.7 c	5.9 a	30.0 a
	Fructification stage (45 dat)			Fructification stage (45 dat)		
	P	K	Mg	Mn	Mo	Zn
1	2.3 b	24.4 b	3.7 a	118.3 a	1.6 b	18.6 c
2	2.8 a	27.4 a	3.1 b	112.0 ab	4.8 a	26.9 b
3	2.8 a	27.2 a	2.8 b	94.7 b	4.3 a	31.5 a
	Harvesting stage (57 dat)			Harvesting stage (57 dat)		
	Ca		Mg		Zn	
1	14.3 a		3.3 a		18.1 b	
2	12.0 b		2.5 b		24.7 a	
3	11.6 b		2.1 b		28.1 a	

Abbreviations: dat, days after transplant; EC, electrical conductivity. Different letters indicate statistical difference (Tukey, $p \leq 0.05$).

For plant growth with 2 dS m^{-1} EC, in the vegetative stage, the concentration of Ca was higher than 1 dS m^{-1} EC by 14.8 %; for fructification, the concentration of P, K, and Mo was higher by 23.8, 12.36, and 200% regarding 1 dS m^{-1} EC, respectively.

The plant growth with 3 dS m^{-1} EC presented higher values in the vegetative stage, with a higher concentration compared to 1 dS m^{-1} EC for P, 14.8%, and for Fe, 22.8%. In the flowering stage with this EC compared to the lowest EC, the highest concentration of N and Zn was obtained, standing out in 12.8 and 19.1%, respectively; for Mo, in this stage, a higher concentration was registered concerning 2 dS m^{-1} EC, higher by 57.9%. In fructification and harvest, for Zn, the highest concentration was presented with 3 dS m^{-1} EC, with the highest value in fructification, surpassing 1 dS m^{-1} EC by 40.9% (Table 1).

For SPAD readings, statistical differences were obtained in the fructification stage, registering the highest value with 2 dS m^{-1} EC (61.57), higher by 5.5% than 1 dS m^{-1} EC (Table S1).

2.5. Sps × EC Combination

For the vegetative stage in the Sps x EC combination, an EC of 3 dS m^{-1} favored the P concentration in *J. procumbens*, for Ca and Fe in *J. tlaxcala* (Figure 3). The plant growth with 2 dS m^{-1} EC favored the P and Mg concentration in *J. tlaxcala*; this same EC level favored K, Ca, and Fe concentration in *J. procumbens*, while with 1 dS m^{-1} EC the highest value was registered for K and Mg, in *J. tlaxcala* and *J. procumbens*, respectively.

In the flowering stage, the plant growth with 3 dS m^{-1} EC favored the concentration of N, P, and Mo in both Sps. With 1 dS m^{-1} EC in both Sps, the highest concentration of Mg and Mn was registered. For B, 1 dS m^{-1} EC favored its concentration in *J. procumbens*, while in *J. tlaxcala* the highest value was registered with 3 dS m^{-1} EC (Figure 4).

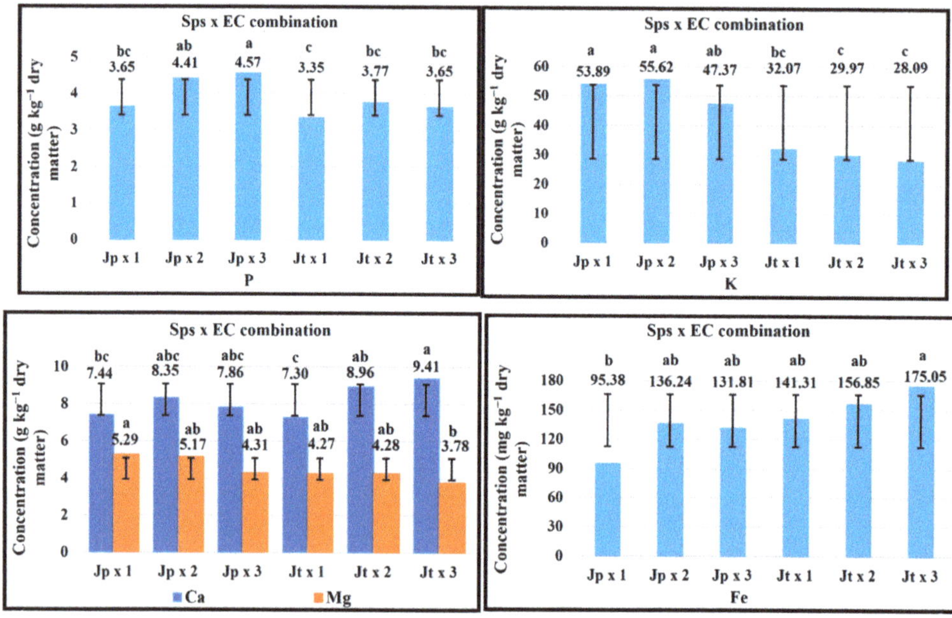

Figure 3. Effect of the Sps x EC combination on the concentration of K, P, Ca, Mg, and Fe in vegetative stage of *Jaltomata procumbens* and *Jaltomata tlaxcala*. Abbreviations: Jp $\times$ 1, *J. procumbens* x 1 dS m^{-1}; Jp $\times$ 2, *J. procumbens* $\times$ 2 dS m^{-1}; Jp $\times$ 3, *J. procumbens* $\times$ 3 dS m^{-1}; Jt $\times$ 1, *J. tlaxcala* $\times$ 1 dS m^{-1}; Jt $\times$ 2, *J. tlaxcala* $\times$ 2 dS m^{-1}; Jt $\times$ 3, *J. tlaxcala* $\times$ 3 dS m^{-1}. Colegio de Postgraduados, Campus Montecillo, Texcoco, State of Mexico. September 2020–February 2021. Data taken between 17 and 129 dat. Different letters among bars indicate statistical difference (Tukey, $p \leq 0.05$).

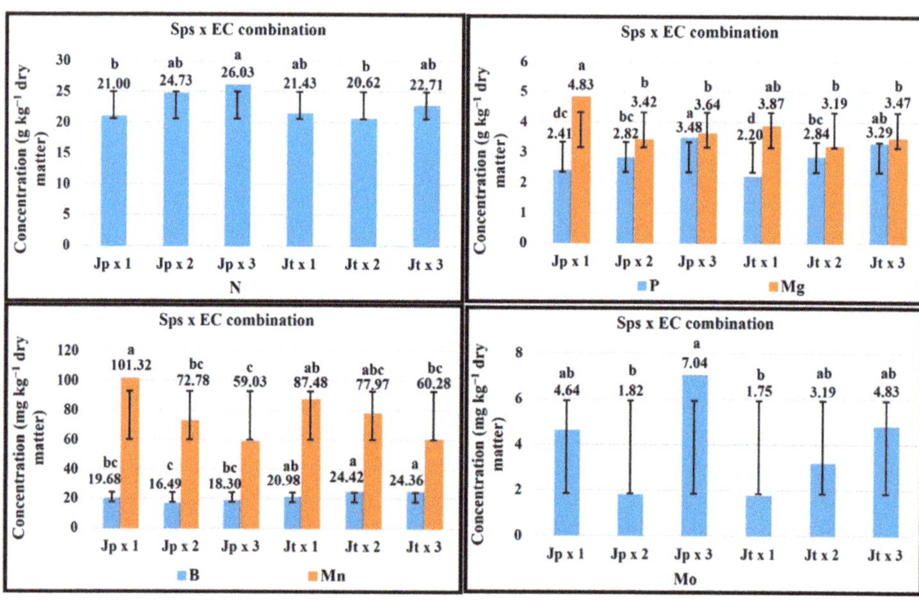

Figure 4. Effect of the Sps x EC combination on the concentration of N, P, Mg, Bo, Mn, and Mo in flowering stage of *Jaltomata procumbens* and *Jaltomata tlaxcala*. Abbreviations: Jp $\times$ 1, *J. procumbens* $\times$ 1 dS m^{-1};

Jp × 2, *J. procumbens* x 2 dS m^{-1}; Jp × 3, *J. procumbens* × 3 dS m^{-1}; Jt × 1, *J. tlaxcala* × 1 dS m^{-1}; Jt × 2, *J. tlaxcala* × 2 dS m^{-1}; Jt × 3, *J. tlaxcala* × 3 dS m^{-1}. Colegio de Postgraduados, Campus Montecillo, Texcoco, State of Mexico. September 2020–February 2021. Data taken between 17 and 129 dat. Different letters among bars indicate statistical difference (Tukey, $p \leq 0.05$).

For fructification, 1 dS m^{-1} EC favored the Ca and Mg concentration in *J. procumbens* and Mg in *J. tlaxcala*. With 2 dS m^{-1}, the K concentration was favored in *J. procumbens*, while for *J. tlaxcala* the P and B concentration, with SPAD readings (63.17). For *J. procumbens*, with 3 dS m^{-1} EC the concentration of P, B, and Zn was promoted, with the SPAD readings (63.23), while, in *J. tlaxcala*, it favored K, Ca, and Zn concentration (Figure 5).

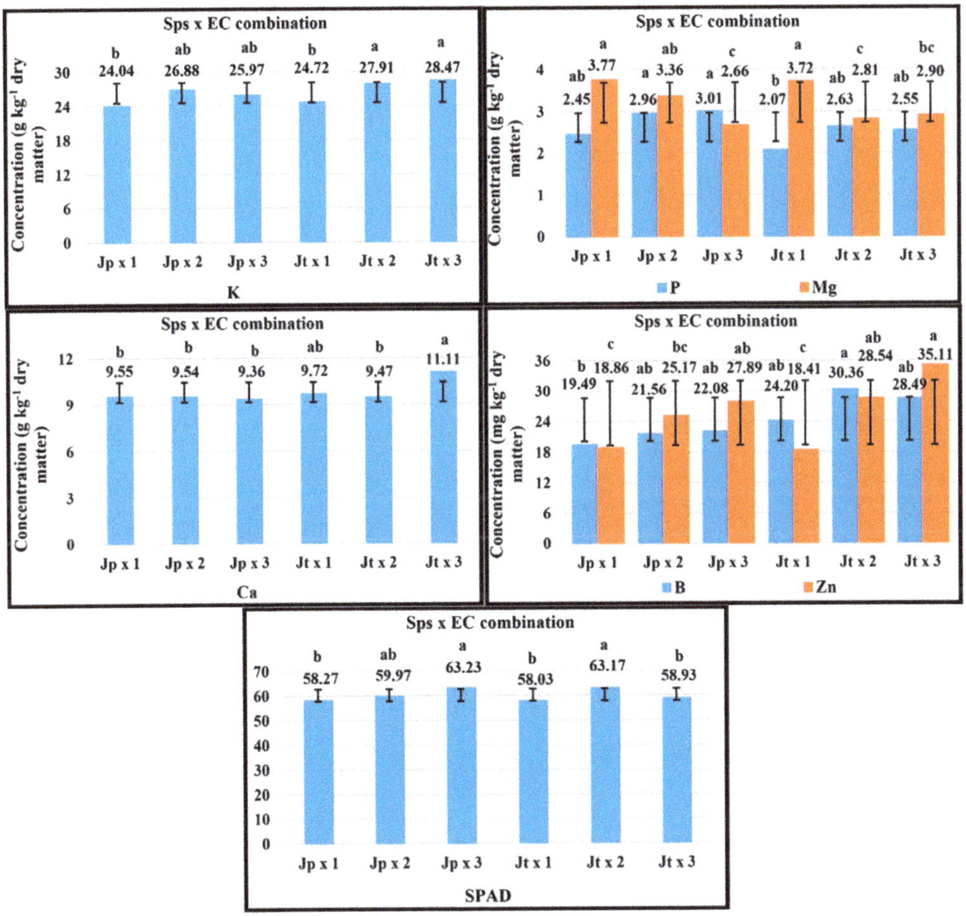

Figure 5. Effect of the Sps x EC combination on the concentration of K, P, Ca, Mg, B, and Zn and SPAD readings, in fructification stage of *Jaltomata procumbens* and *Jaltomata tlaxcala*. Abbreviations: Jp × 1, *J. procumbens* x 1 dS m^{-1}; Jp × 2, *J. procumbens* × 2 dS m^{-1}; Jp × 3, *J. procumbens* × 3 dS m^{-1}; Jt × 1, *J. tlaxcala* × 1 dS m^{-1}; Jt × 2, *J. tlaxcala* × 2 dS m^{-1}; Jt × 3, *J. tlaxcala* × 3 dS m^{-1}. Colegio de Postgraduados, Campus Montecillo, Texcoco, State of Mexico. September 2020– February 2021. Data taken between 17 and 129 dat. Different letters among bars indicate statistical difference (Tukey, $p \leq 0.05$).

For the harvest stage, the plant growth with 1 dS m^{-1} EC favored the concentration of K, Ca, and Mg in *J. procumbens*, and in *J. tlaxcala* for Ca and Mg. With 2 dS m^{-1} EC, in *J. procumbens* the concentration of N was favored, and in *J. tlaxcala* for K and Mn. Finally, 3 dS m^{-1} EC helped a higher concentration of Mn and Zn in *J. procumbens*, while in *J. tlaxcala* in N and Zn (Figure 6).

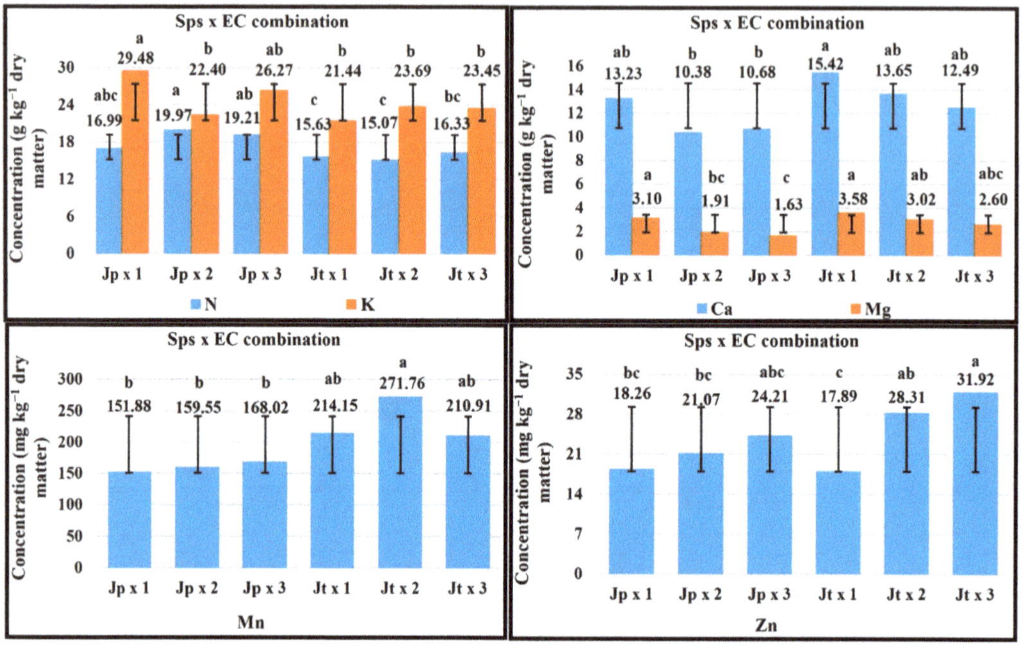

Figure 6. Effect of the Sps x EC combination on the concentration of the N, K, Ca, Mg, Mn, and Zn, in harvest stage of *Jaltomata procumbens* and *Jaltomata tlaxcala*. Abbreviations: Jp × 1, *J. procumbens* × 1 dS m^{-1}; Jp × 2, *J. procumbens* × 2 dS m^{-1}; Jp × 3, *J. procumbens* × 3 dS m^{-1}; Jt × 1, *J. tlaxcala* × 1 dS m^{-1}; Jt × 2, *J. tlaxcala* × 2 dS m^{-1}; Jt × 3, *J. tlaxcala* × 3 dS m^{-1}. Colegio de Postgraduados, Campus Montecillo, Texcoco, State of Mexico. September 2020–February 2021. Data taken between 17 and 129 dat. Different letters among bars indicate statistical difference (Tukey, $p \leq 0.05$).

2.6. Order of Macronutrients and Micronutrients Uptake

The macronutrients uptake in *J. tlaxcala* for the harvesting stage and, in both Sps, for the vegetative, flowering, and fructification stages was K > N > Ca > Mg > P, while for *J. procumbens* at harvesting was K > N > Ca > P > Mg. In the micronutrients, for both Sps, the order was Fe > Mn > B > Zn > Mo in the vegetative stage and, in flowering, Fe > Mn > Zn > B > Mo; in the fructification stage for *J. procumbens* and *J. tlaxcala*, they were Fe > Mn > Zn > B > Mo and Fe > Mn > B > Zn > Mo, respectively; in the harvesting stage for *J. procumbens*, it was Fe > Mn > B > Zn > Mo, and, for *J. tlaxcala*, Mn > Fe > B > Zn > Mo (Figure 7).

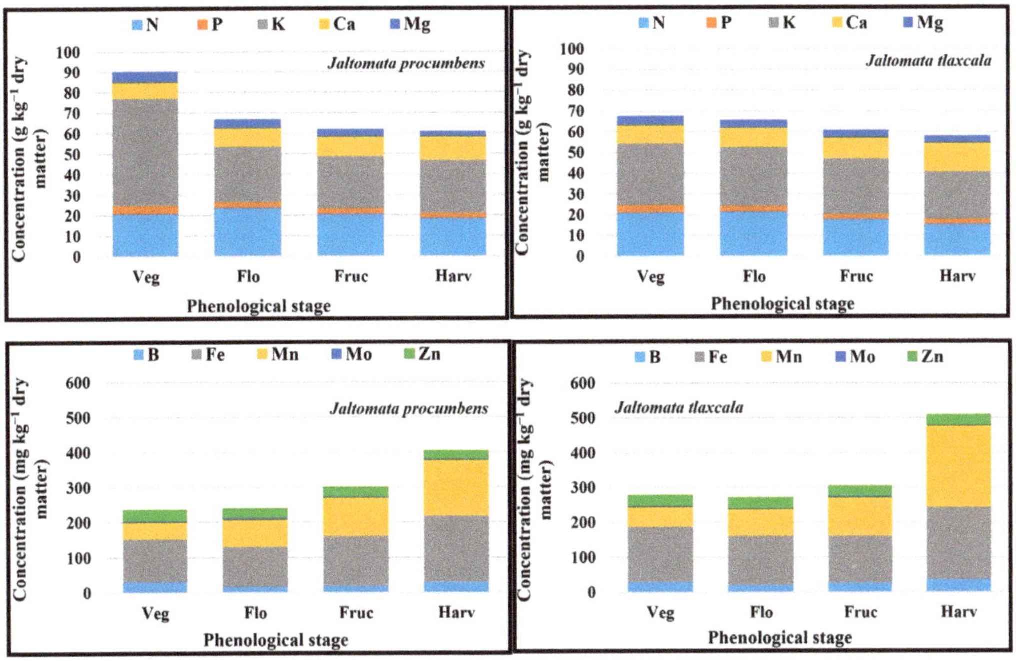

Figure 7. Order of the uptake level on the macronutrients and micronutrients of *Jaltomata procumbens* and *Jaltomata tlaxcala*. Colegio de Postgraduados, Campus Montecillo, Texcoco, State of Mexico. September 2020–February 2021. Sampling carried out at 17, 31, 45, and 57 dat, corresponding to the stages: Veg, Vegetative; Flo, Flowering; Fruc, Fructification; Harv, Harvesting.

2.7. Total Dry Matter

The *J. procumbens* species compared to *J. tlaxcala* had a higher TDM for the vegetative, flowering, fructification, and harvesting stages by 35.5, 61.0, 97.4, and 162.7%, respectively. For EC, in flowering with 3 dS m^{-1}, higher values were registered with respect to 1 dS m^{-1} by 65.2%; for the vegetative, fructification, and harvesting stages, there were no statistical differences, but the same trend was recorded: the higher EC, the higher TDM (Table 2).

Table 2. Comparison of total dry matter means of *Jaltomata procumbens* and *Jaltomata tlaxcala*, cultivated at three electrical conductivity levels of nutritive solution. Colegio de Postgraduados, Campus Montecillo, Texcoco, State of Mexico. September 2020–February 2021. Data taken between 17 and 129 dat.

FV	Total Dry Matter (g plant^{-1})			
Species	Vegetative	Flowering	Fructification	Harvesting
Jaltomata procumbens	3.1 a	19.6 a	55.2 a	230.6 a
Jaltomata tlaxcala	2.3 b	12.1 b	28.0 b	87.8 b
EC (dS m^{-1})				
1	2.4 a	11.9 b	34.2 a	147.1 a
2	2.7 a	15.9 ab	38.2 a	164.5 a
3	3.0 a	19.7 a	52.4 a	165.9 a

Abbreviation: EC, electrical conductivity. Different letters indicate statistical differences (Tukey, $p \leq 0.05$).

2.8. Sps x EC Combination

The Sps x EC combination showed that the plant growth of both Sps with 3 dS m^{-1} EC favored the TDM in the vegetative stage (*J. procumbens* = 3.4 g plant^{-1} and *J. tlaxcala* = 2.7 *g* plant^{-1}) (Table S2), and in flowering, for *J. procumbens* in fructification and for *J. tlaxcala* in harvesting (Figure 8). For *J. tlaxcala* and *J. procumbens*, 2 dS m^{-1} EC favored them in the fructification and harvest stages.

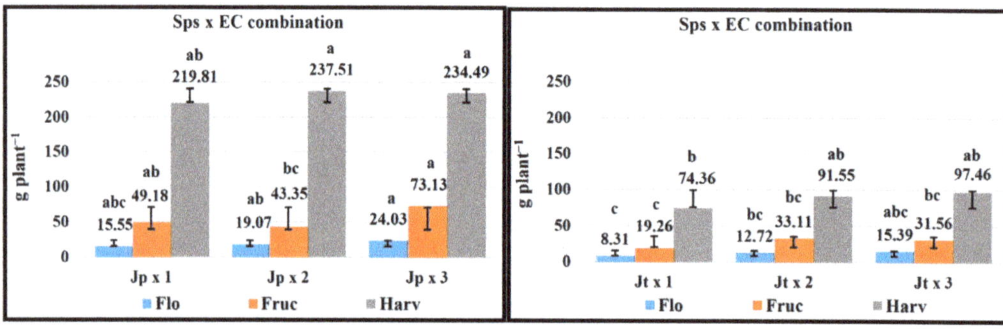

Figure 8. Effect of the Sps × EC combination on the total dry matter of *Jaltomata procumbens* and *Jaltomata tlaxcala*. Abbreviations: Jp × 1, *J. procumbens* × 1 dS m^{-1}; Jp × 2, *J. procumbens* × 2 dS m^{-1}; Jp × 3, *J. procumbens* × 3 dS m^{-1}; Jt × 1, *J. tlaxcala* × 1 dS m^{-1}; Jt × 2, *J. tlaxcala* × 2 dS m^{-1}; Jt × 3, *J. tlaxcala* × 3 dS m^{-1}. Colegio de Postgraduados, Campus Montecillo, Texcoco, State of Mexico. September 2020–February 2021. Data taken between 17 and 129 dat. Different letters among bars indicate statistical difference (Tukey, $p \leq 0.05$).

2.9. Total Dry Matter Accumulation Order by Plant Structure

The distribution of dry matter in leaf (L), root (R), stem (S), and reproductive structures (RE) in *J. procumbens* was in the order of L > R > S > RE in the vegetative and flowering stages, R > S > L > RE for fructification, and RE > R > S > L for harvesting. In *J. tlaxcala*, except for the accumulation of TDM in the vegetative stage which was equal to *J. procumbens* (L > R > S > RE), for the following three stages, a greater accumulation of TDM was registered in root and leaf (flowering = R > L > S > RE and fructification = R > L > RE > S) and root and reproductive structures (harvest = R > RE > L > S) (Figure 9).

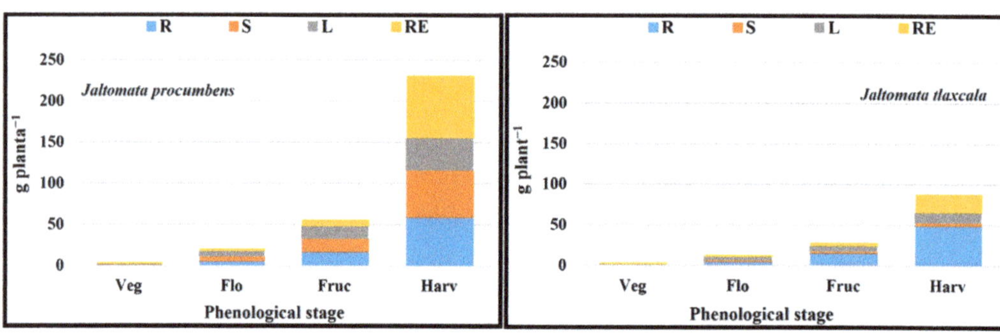

Figure 9. Dry matter accumulation by leaf (L), stem (S), root (R), and reproductive structures (RE) in the vegetative (Veg), flowering (Flo), fructification (Fruc), and harvesting (Harv) stages of *Jaltomata procumbens* and *Jaltomata tlaxcala*. Colegio de Postgraduados, Campus Montecillo, Texcoco, State of Mexico. September 2020–February 2021. Data taken between 17 and 129 dat.

3. Discussion

Variability in the macro- and micronutrients' uptake was observed in both *Jaltomata* species. For macronutrients, *J. procumbens* absorbed a greater amount of N and P, while *J. tlaxcala* did so for Ca. N is involved in vegetative development. It is the most abundant macronutrient in the plant tissues, from 10 to 50 g kg^{-1} of total dry matter [18]. Its function is essential for the plant to complete its biological cycle [19]; it increases the photosynthetic production and duration in the leaf area, which are important characteristics because a larger leaf area is also a determinant of higher yields [20]; therefore, the greater uptake of this element is explained in the early stages of growth, where vegetative development is privileged, reducing uptake during fructification and harvesting to favor fruit development; the behavior of this is shown in both Sps in Figure 1. The N uptake rate is often positively correlated with K uptake, which is an element involved in cell expansion and stomatal opening, enzymatic activation, and osmotic adjustment, contributes up to 10% of the plant's dry weight, and is the most abundant cation in plant cells [21]. This correlation is probably due to charge balance [19] or by the activation of the transporter involved in NO_3^- and K^+ assimilation, since recent works suggest that the NRT1.5 transporter, which directs the transport of NO_3^- from the root to the stem, is also involved in the nitrate-dependent potassium translocation [22], and it is the main counterion for the translocation of NO_3^- to the phloem [21].

Phosphorus had the same uptake dynamics in both Sps and constitutes between 0.5–5 g kg^{-1} of the dry weight of the plant, and the vacuole is the main storage site and plays an important role in cell division and carbohydrate metabolism [23], in the phosphorylation process of the NO_3^- transporters, and in the low- and high-affinity transport systems NRT1 and NRT2 which regulate the absorption of NO_3^- [22,24].

For Ca, the uptake rate and accumulation depend on the species [25]; however, the behavior observed in this work showed that both Sps had the same uptake dynamics (Figure 1). This element plays a structural and signaling role, and it can contribute to the osmotic balance; its concentration in plant tissue varies according to the crop conditions, organ and age of the plant, and the species [26]. This depends on the transpiration rate and water movement through the plant, and on the cationic exchange capacity of the cell wall, which influences the transport of Ca in the apoplast and its storage within the different tissues [25]; Ca is important for cell wall resistance and to prevent its enzymatic degradation, by contributing to the formation of bridges between pectins, resulting in Ca–pectate complexes, keeping fruit firmness and increasing their shelf life [27]. The relative distribution of total Ca is from 70 to 90% in leaves and from 10 to 30% in roots; its concentration in mature leaves can reach 10% of dry weight [26].

The average results of Mg in leaves were 3.5 and 3.6 g kg^{-1} for *J. tlaxcala* and *J. procumbens*, respectively, being in the upper limit of the optimal interval, 1.5–3.5 g kg^{-1}, required in vegetative parts for plant growth [28], with the highest value in *J. tlaxcala*. Mg is important in the chlorophyll formation, it has a key role in the photosynthesis [29]; about 75% of Mg in the leaf is involved in protein synthesis and from 15 to 20% with chlorophyll pigments [28], because it is part of the complex of this molecule [30], whereby, it is preferably transported to this structure, and it also participates in the carbohydrate transport from the source organ to the demand organ [29]. This could respond to the greater absorption of Mg by *J. tlaxcala* at the fructification stage, as this element is required for the carbohydrate transport to the fruit, which represents the main organ of demand; for this Sps, a sucrose content in fruit of 9.4% is reported, while for *J. procumbens* it is 1.6% [10].

For micronutrients in both Sps, the B and Fe uptake presented a similar behavior, registering its lowest point in the flowering stage, and then its content increased. In both

Sps, B registered its lowest point in flowering, which could indicate a greater distribution towards the flowers, where it plays a significant role in their development; subsequently, its concentration in the leaf increases, probably due to their greater requirement for the mobilization of sugars to the fruits. The B requirement in plants correlates with the pectin content [31], which could indicate a higher content of this polysaccharide in *J. tlaxcala*, by registering a higher concentration of B, than *J. procumbens*. The B plays an important role in the development of flowers, seeds, and fruit and participates in the transport of water, nutrients, and sugar towards growing regions; it is considered mobile through the phloem for many agricultural crops and transported as a complex with polyols; it is reported in plants from 10 to 200 mg kg^{-1} [32].

The Fe uptake depends on the species, genotype, environment, and age of the plant [33]; in plants, 140 mg kg^{-1} of dry weight is reported [34] and is stored in the vacuole where transporters such as VIT1 and IREG2 are in charge of importing the element into this organelle [33], where later it is removed to other organelles via the phloem, being the protein of the OPT3 oligopeptide transporter family, identified in Arabidopsis, the one involved in this process [35]. The Fe participates in the synthesis of DNA and chlorophyll and the reduction in N [36]; it is required for processes such as respiration and photosynthesis, by participating in the electron transport chain; it is a constituent of the heme portion of antioxidant enzymes such as catalases (CAT), non-specific peroxidases (POD), ascorbate peroxidases (APX), and as a cofactor of the Fe-superoxide dismutase [33].

Mn and Zn participate in the processes of photosynthesis and protein synthesis [37,38]. Mn also participates in respiration, enzyme activation, and synthesis of fatty acids, with a requirement in the plant of 20–40 mg kg^{-1} of dry weight [37] and is present in all the cellular compartments, being the vacuole which plays a role as temporary storage for its subsequent distribution to the other organelles, and its uptake and distribution is mediated by multiple transport proteins from diverse gene families [39]; on the other hand, Zn participates in other diverse physiological functions such as membrane structure and stress and diseases tolerance; its presence in plants is 30–100 mg kg^{-1} of dry matter [37].

For Mo, both Sps exceeded the requirement of this micronutrient (0.1–1.0 mg kg^{-1}) with values for *J. procumbens* and *J. tlaxcala* of up to 4.5 and 3.65 mg kg^{-1}, respectively. For some species with levels higher than 1.5 mg kg^{-1}, anatomical alterations in the leaf, root, and stem are present; furthermore, it is reported that plants can tolerate high levels of up to 1000 mg kg^{-1} [34]. In this work, no negative effects on the plant development were observed; however, anatomical and physiological studies are required to determine the mechanisms of molybdenum sequestration and tolerance in both Sps at high levels of Mo uptake, as reported in *Brassica* species [40]. This element is important in the function of more than 50 different enzymes such as nitrogenase and nitrate reductase; the process of absorption and redistribution in the plant is still not entirely clear [41].

The effects of EC on the macro- and micronutrients' uptake were observed. In elements such as Mg and Mn, the higher EC, the lower uptake of these nutrients was recorded, while the others, except for Ca in the last stage, showed an increase. This may respond to the fact that many plants have mechanisms to eliminate salts or to support their presence within their cells; so, high saline concentrations that could be harmful to some species are not harmful to others [42], which could be indicating a saline stress tolerance, involving an activation of gene modulations [43].

For SPAD readings, although the highest value was recorded at 2 dS m^{-1} EC, except for the vegetative stage, the highest SPAD readings were presented as EC increased, indicating a greater chlorophyll presence, because these readings are proportional to the concentration of this pigment in the leaf, where the available nutrients are important [7]. High chlorophyll concentrations have been found under conditions of high EC levels, reporting that the increase in chlorophyll under saline conditions improves stress tolerance [44].

For the dry matter accumulation in both Sps, the same dynamics were registered, with the greatest TDM increment for both Sps being from 45 dat, corresponding to the fructification stage (Figure 10), with a greater accumulation in *J. procumbens*, which is attributed to the characteristics of them, because *J. procumbens* presents a greater vegetative development with a higher plant height, thicker stem diameter, more ramifications, and more number of leaves [7]. In this work, *J. procumbens* exceeded *J. tlaxcala* in stem length (SL), root length (RL), stem diameter (SD), and leaf area (LA) in all the sampling stages (Table S3), with values for the last sampling being the following: SL = 162.19 cm, RL = 53.73 cm, SD = 12.83 mm, and LA = 5980.20 cm^2 plant^{-1}, compared with *J. tlaxcala* which registered 64.36 cm for SL, 47.49 cm for RL, 7.91 mm for SD, and 2058.20 cm^2 plant^{-1} for LA.

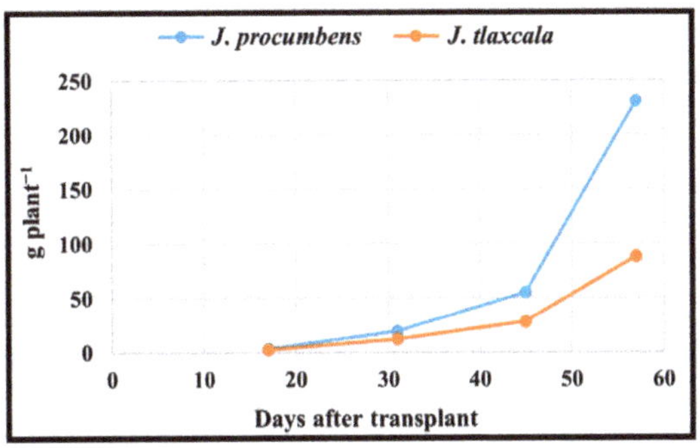

Figure 10. Dynamics of dry matter accumulation of *Jaltomata procumbens* and *Jaltomata tlaxcala*. Colegio de Postgraduados, Campus Montecillo, Texcoco, State of Mexico. September 2020–February 2021. Data taken between 17 and 129 dat.

The importance of dry matter distribution is due to the yield of a crop, which is given by its ability to accumulate biomass in the organs destined for harvest [45].

In *J. procumbens*, the distribution of the TDM (vegetative and flowering = L > R > S > RE; fructification = R > S > L > RE; and harvest = RE > R > S > L) would indicate that, in the early stages, this Sps focuses on the production of vegetative material, to generate the photosynthates that will be stored, and later translocated to the fruits, which represent the main sites of demand for the harvest period, competing among themselves and with the vegetative structures by available photoassimilates [45]. On the contrary, the distribution of TDM in *J. tlaxcala* (vegetative = L > R > S > RE; flowering = R > L > S > RE; fructification = R > L > RE > S; and harvest = R > RE > L > S) would indicate that this Sps spent less time accumulating dry matter in storage structures, allocating to the photoassimilates generated to supply the demand in the fruit development. This is due to the nature of both genotypes because *J. procumbens* presents a continuous and superior generation of ramifications, inflorescences, and fruits than *J. tlaxcala*; therefore, it requires a greater accumulation of reserves to supply photoassimilates to the sites with the greatest demand in the fructification and harvesting stages, which also explains the smaller size and weight of the fruit reported for *J. procumbens* [7], since the greater number of fruits per plant, the smaller the dry matter partition to each of these [46].

As mentioned previously, these species are found in traditional food production systems, with particular environmental and soil conditions depending on the site where they are established. Therefore, it is important to use this information to carry out fertilization to first approximate the different soil conditions where these species grow naturally and what strategies can be applied to maintain their populations.

4. Materials and Methods

The research was carried out in a greenhouse at the Colegio de Postgraduados, Campus Montecillo, Texcoco, State of Mexico, from August to November 2020. Water and nutrients were supplied through drip irrigation without nutrient-solution reuse.

4.1. Genetic Material

The species *Jaltomata procumbens* of an erect growth habit and blue dark fruit and *Jaltomata tlaxcala* of a decumbent growth with green fruit in the mature stage, from Tlaxcala, Mexico (19° 15′ N and 97° 53′ W, at 2500 m of altitude), were evaluated; the region has a predominant temperate subhumid climate C(w1) and C(w2) and an average annual temperature between 12 °C and 18 °C. Seeds of the fruits obtained from a previous work in 2019 were used. The seeds were extracted and rinsed off with tap water, they were dried under shadow at room temperature (22 °C), and they were stored in paper bags of 125 g capacity and maintained in refrigerated conditions at 4 °C.

4.2. Experiment Management

The seeds were placed in petri dishes and were soaked with distilled water for four days and maintained in a germination chamber (model: ATTGPT-B; series: 143958201) with diurnal (12 h at 30 °C $\pm$ 1) and nocturnal (12 h at 20 °C $\pm$ 1) conditions. When necessary, distilled water was used to keep the seeds under soaking conditions.

After soaking time, the seeds were extracted and placed in a sieve (mesh number 22) and rinsed with distilled water for 10 s. The seeds were distributed on moistened filter paper in petri dishes and kept in the germination chamber under the aforementioned conditions. Distilled water was applied to keep the paper moistened.

The germinated seeds, with cotyledonous leaves, were put in plastic cups of 24 mL capacity with peat moss as a substrate and irrigated with distilled water. They were kept for two days in the germination chamber; after that, these were placed in a greenhouse with 40% shade cloth and irrigated with tap water (pH 7.7 and 0.5 dS m^{-1} EC). Then, the plants were transplanted in 265 mL Styrofoam cups with red porous gravel locally known as tezontle as a substrate (particle diameter of 1 to 3 mm) and irrigated with a Steiner nutrient solution with 1 dS m^{-1} EC until they reached a high of 15 cm. Forty-two days after sowing, the plants were transplanted to 40 $\times$ 40 black polyethylene bags (13 L) with red tezontle as a substrate (diameter of $\leq$12 mm, 0.78 g cm^{-3} density, and 21.9% humidity retention). The main stem and branches were moored during the plant growing. In *J. procumbens*, lateral and basal pruning were performed, and in *J. Tlaxcala*, two basal shoots were kept, and the lateral shoots were eliminated. Pests and pathogens, such as beet armyworm (*Spodoptera exigua*) and powdery mildew (*Oidium* sp.), were presented. Beet armyworms were eliminated manually, whereas for powdery mildew, sodium bicarbonate (20 g L^{-1}) was applied.

Four destructive samples were taken at 17, 31, 45 and 57 dat: vegetative, flowering, fructification, and harvesting stages, respectively. The sampled plants were extracted from the substrate and the root was washed with tap water, and at the laboratory the root, leaves, stem, and reproductive structures were separated. To obtain the dry weights, each sample was placed in paper bags and dried in a stove for 72 h (Riossa Stove, model HCF-125D, series 120209).

4.3. Experimental Design

A factorial design was established with the next factors and levels: two species (*J. procumbens* and *J. tlaxcala*), three concentrations (EC of nutrient solution) of Steiner nutrient solution [47], with macro- and micronutrients supplied proportionally to each EC level: 1, 2, and 3 dS m^{-1}, dosed from the transplant; the irrigation volume applied daily was from 0.23 to 1 L per plant, depending on the growth stage. The pH was between a 5.5 to 6.5 interval. The pH and EC were monitored using a portable equipment (Conductronic PC18). The treatments were distributed in a completely randomized design with three replications. The experimental unit was one plant per pot, with 72 experimental units in total.

4.4. Variables Evaluated

4.4.1. Macro- and Micronutrients Analysis in Leaves

The dried leaf samples were milled and passed through a mesh number 40. Then 0.25 g was weighed, and a humid digestion was carried out using a mix of H_2SO_4 and $HClO_4$ (2:1, *v*/*v*), with 1 mL of H_2O_2 at 30%. After that, each sample was diluted to 25 mL with deionized water and then filtered. Except for N, all the nutrients were determined with a coupled plasma atomic emission spectrometry equipment (ICP-AES 725-ES, Agilent, Santa Clara, CA, USA). The N were determined with the Semimicro-Kjeldahl method [48]. The results are reported in g kg^{-1} for macronutrients and in mg kg^{-1} for micronutrients.

4.4.2. Plant Characteristics

Before each sampling, the SPAD-502 readings were used to measure the intensity of the green color of the leaves. An average value of four recently expanded opposite leaves was obtained.

Stem height (SH) and root length (RL) were registered with a tape measure in cm, the stem diameter (SD) with a vernier caliper in mm, and the leaf area (LA) with a leaf area integrator in cm^2 (LI-COR, Inc. Lincoln, NE, USA, LI-3100 Area meter). The total dry matter (TDM) was obtained with the sum of the dry weights of each organ and was measured with an analytical balance in g (Adventurer Pro AV213C Model).

4.4.3. Statistical Analysis

A variance analysis was performed based on a completely random factorial model 2×3 and a Tukey's separation means test ($p \leq 0.05$). The SAS program (2008) 9.2 version statistical program was used [49].

5. Conclusions

A variability was registered between the species and phenological stages, in the macro- and micronutrients concentration, and in the total dry matter accumulation.

The *Jaltomata procumbens* absorbed a higher amount of N, P, and Mo, while *Jaltomata tlaxcala* absorbed more Ca, B, Fe, and Zn.

The plant growth of both species with high electrical conductivity favored the total dry matter accumulation.

Due to this variability, specific fertilization programs are required for each species. The optimal nutrient absorption and biomass production are obtained with the nutrient solutions with 2 and 3 dS m^{-1} EC, for *Jaltomata procumbens* and *Jaltomata tlaxcala*, respectively, in a greenhouse and using hydroponics.

It is necessary to use these data to carry out fertilization to first approximate the field conditions due to the different soil characteristics where these species grow.

Supplementary Materials: The following supporting information can be downloaded at: https://www.mdpi.com/article/10.3390/plants14071124/s1, Table S1: SPAD readings of *Jaltomata procumbens* and *Jaltomata tlaxcala*, cultivated at three electrical conductivity levels of nutritive solution. Colegio de Postgraduados, Campus Montecillo, Texcoco, State of Mexico. September 2020–February 2021. Data taken at 17, 31, 45, and 57 dat, corresponding to vegetative, flowering, fructification, and harvesting stages, respectively; Table S2: Effect of the Sps x EC combination on the total dry matter of *Jaltomata procumbens* and *Jaltomata tlaxcala* in the vegetative stage. Colegio de Postgraduados, Campus Montecillo, Texcoco, State of Mexico. September 2020–February 2021. Data taken between 17 and 129 dat; Table S3: Effect of the species on the plant structures of *Jaltomata procumbens* and *Jaltomata tlaxcala*. Colegio de Postgraduados, Campus Montecillo, Texcoco, State of Mexico. September 2020–February 2021. Data taken between 17 and 129 dat.

Author Contributions: Conceptualization, I.D.F.-S.; Formal analysis, I.D.F.-S.; Investigation, I.D.F.-S.; Methodology, I.D.F.-S.; Resources, M.S.-V. and E.U.-M.; Supervision, M.S.-V.; Writing—original draft, I.D.F.-S. and M.S.-V.; Writing—review and editing, I.D.F.-S., M.S.-V. and E.U.-M. All authors have read and agreed to the published version of the manuscript.

Funding: This research was funded by the scholarship program of the Consejo Nacional de Ciencia, Humanidades y Tecnología of Mexico (CONAHCYT) and the Colegio de Postgraduados.

Data Availability Statement: The original contributions presented in this study are included in the article/Supplementary Materials. Further inquiries can be directed to the corresponding author(s).

Acknowledgments: They acknowledge the support of the Consejo Nacional de Ciencia, Humanidades y Tecnología of Mexico (CONAHCYT) for the scholarship given to the first author for his postgraduate studies and of the Colegio de Postgraduados, Campus Montecillo, for the materials provided to carry out this research. They also acknowledge the support of the Libia I. Trejo-Téllez and Guadalupe Peralta- Sánchez (postgraduate in Edaphology, Plant Nutrition Department, Colegio de Postgraduados, Campus Montecillo) during the nutrient determination.

Conflicts of Interest: The authors declare no conflicts of interest.

References

1. González, C.F.; Cabezas, G.M.; Ramírez-Gómez, M.; Ramírez, D.J. Curvas de absorción de macronutrientes en tres variedades de caña de azúcar (*Saccharum officinarum* L.) para panela, en la hoya del río Suárez. *Rev. UDCA Actual. Divulg. Cient.* **2018**, *21*, 395–404. [CrossRef]
2. Gastelum-Osorio, D.A.; Sandoval-Villa, M.; Trejo-López, C.; Castro-Brindis, R. Fuerza iónica de la solución nutritiva y densidad de plantación sobre la producción y calidad de frutos de *Physalis peruviana* L. *Rev. Chapingo Ser. Hortic.* **2013**, *19*, 197–210. [CrossRef]
3. Mione, T.; Leiva, G.S.; Yacher, L. *Jaltomata sanmiguelina* (Solanaceae), a new species from northern Peru. University Proceedings, Volga region. *Nat. Sci.* **2015**, *3*, 3–13.
4. Flores-Sánchez, I.D.; Sandoval-Villa, M.; Uscanga-Mortera, E.; Silva-Rojas, H.V. *Jaltomata tlaxcala*, a new species of the genus *Jaltomata* (Solanaceae, Solanoideae, Solaneae). *Phytotaxa* **2024**, *641*, 85–98. [CrossRef]
5. Fonteyne, S.; Castillo, C.J.B.; Lopez-Ridaura, S.; Van, L.J.; Espidio, B.J.; Osorio, A.L.; Martínez, H.F.; Odjo, S.; Verhulst, N. Review of agronomic research on the milpa, the traditional polyculture system of Mesoamerica. *Front. Agron.* **2023**, *5*, 1115490. [CrossRef]
6. Ortiz, G.S.; Saynes, S.V.; Bunge, V.V.; Anglés-Hernández, M.; Pérez, M.E.; Prado, B. Soil governance and sustainable agriculture in Mexico. *Soil Secur.* **2022**, *7*, 100059. [CrossRef]
7. Flores-Sánchez, I.D.; Sandoval-Villa, M.; Alcántar-González, E.G.; Sánchez-García, P.; Soto-Hernández, R.M.; Uscanga-Mortera, E. Caracteres agronómicos de dos poblaciones de *Jaltomata procumbens* (Cav.) J. L. Gentry. *Agron. Mesoam.* **2021**, *32*, 733–749. [CrossRef]
8. Flores-Sánchez, I.D.; Sandoval-Villa, M.; Soto-Hernández, R.M. Effects of electrical conductivity and pruning on secondary metabolite contents in fruits of *Jaltomata procumbens* (Cav.) J. L. Gentry. *Nat. Prod. Commun.* **2023**, *18*, 1–11. Available online: https://journals.sagepub.com/doi/10.1177/1934578X221150547 (accessed on 12 May 2024). [CrossRef]

9. Mera-Ovando, L.M.; Alvarado-Flores, R.; Basurto-peña, F.; Bye-Bottler, R.; Castro-Lara, D.; Evangelista, V.; Mapes-Sánchez, C.; Martínez-Alfaro, M.Ä.; Molina, N.; Saldívar, J. "De quelites me como un taco". Experiencia en educación nutricional. *Rev. Jardín Botánico Nac.* **2003**, *24*, 45–49. Available online: https://www.jstor.org/stable/42597191 (accessed on 22 June 2021).
10. Williams, D.E. Tres Arvenses Solanáceas Comestibles y su Proceso de Domesticación en el Estado de Tlaxcala, México. Master's Thesis, Colegio de Postgraduados, Montecillo, Mexico, 1985. Available online: http://catalogo.colpos.mx/cgi-bin/koha/opac-detail.pl?biblionumber=16343 (accessed on 7 March 2022).
11. WHO. *Guidance on Mainstreaming Biodiversity for Nutrition and Health*; World Health Organization: Geneva, Switzerland, 2020; pp. 1–100. Available online: https://www.who.int/publications/i/item/9789240006690 (accessed on 18 September 2022).
12. Li, X.; Siddique, K.H.M.; Akinnifesi, F.; Callens, K.; Broca, S.; Noorani, A.; Henrich, G.; Chikelu, M.; Bayasgalanbat, N. Introduction: Setting the scene. In *Future Smart Food—Rediscovering Hidden Treasures of Neglected and Underutilized Species for Zero Hunger in Asia*; Xuan, L., Kadambot, H.M.S., Eds.; FAO: Rome, Italy, 2018; pp. 15–32. Available online: http://www.fao.org/3/I8907EN/i8907en.pdf (accessed on 5 May 2022).
13. FAO. *Neglected and Underutilized Crops Species*; COAG/2018/INF/7; Food and Agriculture Organization of the United Nations: Rome, Italy, 2018; pp. 1–5. Available online: https://openknowledge.fao.org/items/3f2c9e30-d29c-4908-94d0-511fe4ffaeaf (accessed on 25 July 2023).
14. N'Danikou, S.; Tchokponhoue, D.A. Plant domestication for enhanced food security. In *Zero Hunger*; Filho, W.L., Azul, A., Brandli, L., Özuyar, P., Wall, T., Eds.; Springer: Cham, Switzerland, 2019; pp. 1–11. [CrossRef]
15. Li, X.; Yadav, R.; Siddique, K.H.M. Neglected and underutilized crop species: The key to improving dietary diversity and fighting hunger and malnutrition in Asia and the Pacific. *Front. Nutr.* **2020**, *7*, 593711. [CrossRef]
16. Sarukhán, J.; (Universidad Nacional Autónoma de México, Ciudad de México, Mexico). Personal communication, 2020.
17. Castro-Villarreal, J.A.; Villarreal-Núñez, J.E. Absorción de nutrientes en tomate industrial IDIAP–T7 en suelos de la región de Azuero. *Cienc. Agropecu.* **2020**, *30*, 27–44. Available online: http://www.revistacienciaagropecuaria.ac.pa/index.php/ciencia-agropecuaria/article/view/125 (accessed on 12 November 2022).
18. Muratore, C.; Espen, L.; Prinsi, B. Nitrogen uptake in plants: The plasma membrane root transport systems from a physiological and proteomic perspective. *Plants* **2021**, *10*, 681. [CrossRef] [PubMed]
19. Feng, H.; Fan, X.; Miller, A.J.; Xu, G. Plant nitrogen uptake and assimilation: Regulation of cellular pH homeostasis. *J. Exp. Bot.* **2020**, *71*, 4380–4392. [CrossRef]
20. Leghari, S.J.; Wahocho, N.A.; Laghari, G.M.; HafeezLaghari, A.; MustafaBhabhan, G.; HussainTalpur, K.; Bhutto, T.A.; Wahocho, S.A.; Lashari, A.A. Role of nitrogen for plant growth and development: A review. *Adv. Environ. Biol.* **2016**, *10*, 209–219. Available online: https://www.researchgate.net/publication/309704090_Role_of_Nitrogen_for_plant_Growth_and_Development_A_review#:~:text=All%20plants%20utilize%20nitrogen%20(N,and%20physiological%20functions%20of%20plant (accessed on 15 November 2021).
21. Ragel, P.; Raddatz, N.; Leidi, E.O.; Quintero, F.J.; Pardo, J.M. Regulation of K^+ Nutrition in Plants. *Front. Plant Sci.* **2019**, *10*, 281. [CrossRef]
22. Fan, X.; Naz, M.; Fan, X.; Xuan, W.; Miller, A.J.; Xu, G. Plant nitrate transporters: From gene function to application. *J. Exp. Bot.* **2017**, *68*, 2463–2475. [CrossRef]
23. Malhotra, H.; Vandana; Sharma, S.; Pandey, R. Phosphorus nutrition: Plant growth in response to deficiency and excess. In *Plant Nutrients and Abiotic Stress Tolerance*, 1st ed.; Hasanuzzaman, M., Fujita, M., Oku, H., Nahar, K., Hawrylak-Nowak, B., Eds.; Springer: Singapore, 2018; Volume 1, pp. 171–190. [CrossRef]
24. Jacquot, A.; Chaput, V.; Mauries, A.; Li, Z.; Tillard, P.; Fizames, C.; Bonillo, P.; Bellegarde, F.; Laugier, E.; Santoni, V.; et al. NRT2.1 C-terminus phosphorylation prevents root high affinity nitrate uptake activity in Arabidopsis thaliana. *New Phytol.* **2020**, *228*, 1038–1054. [CrossRef]
25. Gilliham, M.; Dayod, M.; Hocking, B.J.; Xu, B.; Conn, S.J.; Kaiser, B.N.; Leigh, R.A.; Tyerman, S.D. Calcium delivery and storage in plant leaves: Exploring the link with water flow. *J. Exp. Bot.* **2011**, *62*, 2233–2250. [CrossRef]
26. González-Fontes, A.; Navarro-Gochicoa, M.T.; Ceacero, C.J.; Herrera-Rodríguez, M.B.; Camacho-Cristóbal, J.J.; Rexach, J. Understanding calcium transport and signaling, and its use efficiency in vascular plants. In *Plant Macronutrient Use Efficiency*, 1st ed.; Hossain, M.A., Kamiya, T., Burrit, D., Tran, L.-S.P., Fujiwara, T., Eds.; Elsevier: London, UK, 2017; Volume 1, pp. 165–180. [CrossRef]
27. Ribeiro, L.R.; Leonel, S.; Azevedo, J.M.S.; García, E.L.; Leonel, M.; Honorato, M.L.N.; Silva, M.S.; Ferreira, R.B. Improving the nutritional value and extending shelf life of red guava by adding calcium chloride. *LWT—Food Sci. Technol.* **2020**, *130*, e109655. [CrossRef]
28. Guo, W.; Nazim, H.; Liang, Z.; Yang, D. Magnesium deficiency in plants: An urgent problem. *Crop J.* **2016**, *4*, 83–91. [CrossRef]

29. Farhat, N.; Elkhouni, A.; Zorrig, W.; Smaoui, A.; Abdelly, C.; Rabhi, M. Effects of magnesium deficiency on photosynthesis and carbohydrate partitioning. *Acta Physiol. Plant.* **2016**, *38*, 145. [CrossRef]
30. Tang, R.-J.; Luan, S. Regulation of calcium and magnesium homeostasis in plants: From transporters to signaling network. *Curr. Opin. Plant Biol.* **2017**, *39*, 97–105. [CrossRef] [PubMed]
31. Yoshinari, A.; Takano, J. Insights into the mechanisms underlying boron homeostasis in plants. *Front. Plant Sci.* **2017**, *8*, 1951. [CrossRef]
32. Karthika, K.S.; Rashmi, I.; Parvathi, M.S. Biological functions, uptake, and transport of essential nutrients in relation to plant growth. In *Plant Nutrients and Abiotic Stress Tolerance*, 1st ed.; Hasanuzzaman, M., Fujita, M., Oku, H., Nahar, K., Hawrylak-Nowak, B., Eds.; Springer: Singapore, 2018; Volume 1, pp. 1–49. [CrossRef]
33. Krohling, C.A.; Eutrópio, F.J.; Azevedo, B.A.; Barros, D.L.; Campostrini, E.; Dias, T.; Coutinho, R.A. Ecophysiology of iron homeostasis in plants. *Soil Sci. Plant Nutr.* **2016**, *62*, 39–47. [CrossRef]
34. Anjum, N.A.; Singh, H.P.; Khan, M.I.R.; Masood, A.; Per, T.S.; Negi, A.; Batish, D.R.; Khan, N.A.; Duarte, A.C.; Pereira, E.; et al. Too much is bad—An appraisal of phytotoxicity of elevated plant-beneficial heavy metal ions. *Environ. Sci. Pollut. Res.* **2015**, *22*, 3361–3382. [CrossRef] [PubMed]
35. Connorton, J.M.; Balk, J.; Rodríguez-Celma, J. Iron homeostasis in plants—A brief overview. *Metallomics* **2017**, *9*, 813–823. [CrossRef]
36. Rai, S.; Singh, P.K.; Mankotia, S.; Swain, J.; Satbhai, S.B. Iron homeostasis in plants and its crosstalk with copper, zinc, and manganese. *Plant Stress* **2021**, *1*, 100008. [CrossRef]
37. Li, J.; Jia, Y.; Dong, R.; Huang, R.; Liu, P.; Li, X.; Wang, Z.; Liu, G.; Chen, Z. Advances in the mechanisms of plant tolerance to manganese toxicity. *Int. J. Mol. Sci.* **2019**, *20*, 5096. [CrossRef]
38. Noulas, C.; Tziouvalekas, M.; Karyotis, T. Zinc in soils, water, and food crops. *J. Trace Elem. Med. Biol.* **2018**, *49*, 252–260. [CrossRef]
39. Alejandro, S.; Höller, S.; Meier, B.; Peiter, E. Manganese in plants: From acquisition to subcellular allocation. *Front. Plant Sci.* **2020**, *11*, 300. [CrossRef]
40. Hale, K.L.; McGrath, S.P.; Lombi, E.; Stack, S.M.; Terry, N.; Pickering, I.J.; George, G.N.; Pilon-Smits, E.A.H. Molybdenum sequestration in Brassica species. A role for anthocyanins? *Plant Physiol.* **2001**, *126*, 1391–1402. [CrossRef] [PubMed]
41. Rana, M.S.; Bhantana, P.; Imran, M.; Saleem, M.H.; Moussa, M.G.; Khan, Z.; Khan, I.; Alam, M.D.; Abbas, M.; Binyamin, R.; et al. Molybdenum potential vital role in plants metabolism for optimizing the growth and development. *Ann. Environ. Sci. Toxicol.* **2020**, *4*, 32–44. [CrossRef]
42. Stefanov, M.; Yotsova, E.; Rashkov, G.; Ivanova, K.; Markovska, Y.; Apostolova, E.L. Effects of salinity on the photosynthetic apparatus of two *Paulownia* lines. *Plant Physiol. Biochem.* **2016**, *101*, 54–59. [CrossRef] [PubMed]
43. Muchate, N.S.; Nikalje, G.C.; Rajurkar, N.S.; Suprasanna, P.; Nikam, T.D. Plant salt stress: Adaptive responses, tolerance mechanism and bioengineering for salt tolerance. *Bot. Rev.* **2016**, *82*, 371–406. [CrossRef]
44. Ding, X.; Jiang, Y.; Zhao, H.; Guo, D.; He, L.; Liu, F.; Zhou, Q.; Nandwani, D.; Hui, D.; Yu, J. Electrical conductivity of nutrient solution influenced photosynthesis, quality, and antioxidant enzyme activity of pakchoi (*Brassica campestris* L. ssp. Chinensis) in a hydroponic system. *PLoS ONE* **2018**, *13*, e0202090. [CrossRef]
45. Gandica, O.H.; Peña, H. Acumulación de materia seca y balance de nutrientes en tomate (*Solanum lycopersicum* L.) cultivado en ambiente protegido. *Bioagro* **2015**, *27*, 111–120. Available online: https://ve.scielo.org/pdf/ba/v27n2/art07.pdf (accessed on 8 March 2022).
46. Pérez-Labrada, F.; Benavides-Mendoza, A.; Valdez-Aguilar, L.A.; Robledo-Torres, V. Citric acid in the nutrient solution increases the mineral absorption in potted tomato grown in calcareous soil. *Pak. J. Bot.* **2016**, *48*, 67–74. Available online: https://www.researchgate.net/publication/295086648_Citric_acid_in_the_nutrient_solution_increases_the_mineral_absorption_in_potted_tomato_grown_in_calcareous_soil (accessed on 15 July 2022).
47. Steiner, A.A. The universal nutrient solution. In Proceedings of the Sixth International Congress on Soilless Culture, ISOSC, Lunteren, The Netherlands, 29 April–5 May 1984; pp. 633–650.
48. Horneck, D.A.; Miller, R.O. Determination of total nitrogen in plant tissue. In *Handbook of Reference Methods for Plant Analysis*, 1st ed.; Kalra, Y.P., Ed.; Taylor & Francis Group: New York, NY, USA, 1997; pp. 75–83. Available online: https://scholar.google.com/scholar_lookup?hl=en&publication_year=1998&author=DA+Horneck&author=RO+Miller&&title=Determination+of+total+nitrogen+in+plant+tissue (accessed on 21 February 2025).
49. SAS Institute. *User's Guide Version 9.2*; SAS Institute Inc.: Cary, NC, USA, 2008.

plants

Article

Growth and Biomass Distribution Responses of *Populus tomentosa* to Long-Term Water–Nitrogen Coupling in the North China Plain

Yafei Wang [1,†], Juntao Liu [1,2,†], Yuelin He [3], Wei Zhu [4], Liming Jia [1,*] and Benye Xi [1,*]

[1] Key Laboratory of Silviculture and Conservation of the Ministry of Education, College of Forestry, Beijing Forestry University, Beijing 100083, China; mrw1996@bjfu.edu.cn (Y.W.); ljt1120@bjfu.edu.cn (J.L.)
[2] Guangdong Provincial Key Laboratory of Silviculture, Protection and Utilization/Guangdong Academy of Forestry, Guangzhou 510520, China
[3] National Key Laboratory for Development and Utilization of Forest Food Resources, Zhejiang A&F University, Hangzhou 311300, China; hylhelen@163.com
[4] College of Biological and Environmental Engineering, Jingdezhen University, Jingdezhen 333400, China; 18811375082@163.com
[*] Correspondence: jlm@bjfu.edu.cn (L.J.); benyexi@bjfu.edu.cn (B.X.)
[†] These authors contributed equally to this work.

Abstract: From 2016 to 2021, a field experiment was conducted in the North China Plain to study the long-term effects of drip irrigation and nitrogen coupling on the growth, biomass allocation, and irrigation water and fertilizer use efficiency of short-rotation triploid *Populus tomentosa* plantations. The experiment adopted a completely randomized block design, with one control (CK) and six water–nitrogen coupling treatments (IF, two irrigation levels × three nitrogen application levels). Data analysis was conducted using ANOVA, regression models, Spearman's correlation analysis, and path analysis. The results showed that the effects of water and nitrogen treatments on the annual increment of diameter at breast height (ΔDBH), annual increment of tree height (ΔH), basal area of the stand (BA_S), stand volume (V_S), and annual forest productivity (AFP) in short-rotation forestry exhibited a significant stand age effect. The coupling of water and nitrogen significantly promoted the DBH growth of 2-year-old trees ($p < 0.05$), but after 3 years of age, the promoting effect of water and nitrogen coupling gradually diminished. In the 6th year, the above-ground biomass of *Populus tomentosa* was 5.16 to 6.62 times the under-ground biomass under different treatments. Compared to the I45 treatment (irrigation at soil water potential of −45 kPa), the irrigation water use efficiency of the I20 treatment (−20 kPa) decreased by 88.79%. PFP showed a downward trend with the increase in fertilization amount, dropping by 130.95% and 132.86% under the I20 and I45 irrigation levels. Path analysis indicated that irrigation had a significant effect on the BA_S, V_S, AFP, and TGB of 6-year-old *Populus tomentosa* ($p < 0.05$), with the universality of irrigation being higher than that of fertilization. It is recommended to implement phased water and fertilizer management for *Populus tomentosa* plantations in the North China Plain. During 1–3 years of tree age, adequate irrigation should be ensured and nitrogen fertilizer application increased. Between the ages of 4 and 6, irrigation and fertilization should be ceased to reduce resource wastage. This work provides scientific guidance for water and fertilizer management in short-rotation plantations.

Keywords: short-rotation plantations; tree growth; biomass allocation; irrigation water use efficiency; nitrogen fertilizer partial productivity

1. Introduction

The total amount of natural forest land worldwide declined more sharply, by almost 301 million hectares, over the thirty years between 1990 and 2020. Nonetheless, there was an increase of 123 million hectares in the amount of planted forests. Though the demand for lumber is still growing globally, the amount of timber produced by natural forests worldwide has already reached its peak [1]. Therefore, to resolve the current supply–demand imbalance in the international timber market, the foremost priority is to improve the yield and quality of plantation forests through sustainable management strategies that are effective, scientific, and rational, which requires comprehensive research on the long-term effects of various intensive cultivation methods on the ecological functions of plantations [2–4].

In the present day, the ingenious integration of irrigation and fertilization has become an established developmental path for achieving water–fertilizer integration, as well as a critical pathway for resolving the issues of irrational water and fertilizer application [5,6]. Water and fertilizer coupling has the distinct advantage of leveraging water to improve fertilizer efficacy while also utilizing fertilizer to control water usage. As a result, among the numerous significant measures for intensive plantation management, promoting tree growth through water–fertilizer coupling frequently outperforms the results achieved by implementing irrigation or fertilization alone [7–10]. The overall effect of water and fertilizer synergy is proportional to their ratio and varies with changes in each component. Under situations of adequate soil moisture, the interaction between water and fertilizer can have a superposition effect, promoting plant growth [11–13]. When soil moisture is in a restricted state, excessive application of nitrogen fertilizer not only fails to promote plant growth but also produces antagonistic effects [14,15]. In plantation cultivation, a thorough understanding of tree growth mechanisms is critical for accomplishing long-term forest resource management. The dynamic process of tree growth refers to the patterns of changes in radial growth, height, and overall volume of trees as they mature. Investigating these dynamic processes allows us to understand the underlying laws of stand growth. Understanding tree development patterns under long-term water and nitrogen coupling is very useful for developing dynamic irrigation and fertilization methods during the rotation period [16,17]. According to our understanding, there are still few ongoing studies and publications on the long-term impacts of water and nitrogen [8,9,18].

During the growth and development process, trees prioritize the allocation of biomass to organs capable of acquiring limited resources, adjusting to the environment through constant optimization and modification [19]. A logical biomass allocation scheme is critical for forest tree growth because it can significantly increase growth rate and survival capacity [20]. The logical allocation of above-ground biomass (AGB) can improve light energy usage efficiency, boost photosynthesis, and maximize radial and height development [21,22]. Meanwhile, below-ground biomass (BGB) is essential for plants to compete for limited soil nutrient resources [23,24]. This makes it obvious that the distribution of biomass is intimately linked to the forest tree growth process and is a valuable method and instrument for identifying forest tree growth strategies. Nonetheless, there are still several obstacles in the field of biomass research today. The majority of research has concentrated mostly on the aboveground portions because biomass sampling is very challenging, particularly for belowground biomass. The allocation patterns of various root organs are frequently overlooked, even in studies that focus on the subsurface [22,24,25]. Because of this, we are still unable to evaluate the growth patterns and environmental adaptation of many tree species, especially broad-leaved wood species that have the capacity to develop. Meanwhile, biomass is a fundamental component of forest growth and a key indicator of plantation productivity. The distribution of biomass among tree organs reflects their adaptive methods in growth space. This allocation pattern represents the trade-offs that trees make in the acquisition, accumulation, usage, and distribution of resources [23,26,27]. On the one hand, it is controlled by the species and genotype of trees, and the distribution of biomass may change significantly among tree species [24,25,27,28]. However, according to the functional balance theory and the optimal allocation theory, external environmental factors (like forest age, light, water, and nutrients) also have a significant impact on the patterns of biomass allocation [9,19,29–33]. Therefore, it is crucial to advance the theoretical and practical innovation of plantation cultivation techniques by carrying out a comprehensive study on the biomass allocation patterns of forest trees and resolving the limitations of previous studies.

Poplar (*Populus* spp.) is grown extensively around the world; professional poplar plantations span about 9.4 million hectares, mostly in China, Canada, and Europe. In addition to its many beneficial qualities, such as its propensity for hybridization and ease of propagation [34], it is essential for the world's timber supply, shelterbelt construction, energy provision, carbon sequestration, degraded land restoration, and soil and water conservation [3,34]. In China, poplar plantations account for 6.54% of the timber forest area, yet they contribute to nearly 30% of the country's timber production [3]. Poplars in China are primarily planted in areas with poor site conditions, with the white poplar (*Populus tomentosa*) often cultivated on sandy soils in the Yellow River floodplain region. Water availability is a crucial limiting factor for their growth because they are a species that uses a lot of water [34]. Furthermore, prior research has demonstrated that poplar trees have a greater preference for nitrogen than other fertility elements, making nitrogen the main factor restricting the growth of both trees and crops in northern China [8,9,35]. It appears that extensive water and nitrogen management strategies are crucial for increasing the productivity and quality of poplar plantations in northern China, as well as mitigating the negative impacts of water scarcity on tree development and fitness.

A five-year irrigation and fertilization study was conducted in *Populus tomentosa* plantations in the North China Plain to increase poplar productivity and evaluate the effects of water and nitrogen coupling on plant production. The objectives of this study were to (1) investigate the long-term response of *Populus tomentosa* growth processes to water–nitrogen coupling measures during the rotation period; (2) investigate the allocation patterns of *Populus tomentosa* biomass at the end of the rotation under long-term water–nitrogen coupling; (3) determine whether water–nitrogen coupling could change the water use efficiency and nitrogen fertilizer partial productivity of *Populus tomentosa* over the entire rotation period. This study will serve as a reference for the high-yield, scientific, and sustainable management of worldwide short-rotation plantations.

2. Results

2.1. Long-Term Effects of Different Water–Nitrogen Coupling Treatments on Tree Growth

For individual *Populus tomentosa* trees, compared to the control treatment, the water–nitrogen coupling treatment significantly increased ΔDBH at two years of age, with the I20FH treatment showing the greatest increase, 37.30% higher than the CK treatment ($p < 0.05$). The difference in ΔDBH under different treatments gradually decreased as the trees grew to 3 to 6 years old (Table 1). Throughout the investigation, the water–nitrogen coupling treatment had no significant influence on tree height (Table 1). As stand age increased, ΔDBH and ΔH decreased annually (Figure 1). The difference in irrigation levels was reflected in the DBH increment at 2 years old; the fertilization effect was manifested in the height increment at 5 years old. No significant interaction effect between irrigation and fertilization was observed in the annual DBH and height increments (Figure 1 and Table 1).

Table 1. The one-way ANOVA results of the differences in growth indicators between the water and nitrogen treatment (IF) and the control treatment (CK). The two-way ANOVA assessed the effects of irrigation level, fertilization level, and the interaction between irrigation and fertilization on various growth indicators. In the table, ΔDBH represents the annual increment of individual tree diameter at breast height, ΔH denotes the annual increment of individual tree height, BA_S stands for the stand basal area in different years, Vs indicates the stand volume in different years, and AFP refers to the average forest productivity in different years. Bold values indicate statistically significant effects.

Indices	Source of Variation	Stand Ages (Years Old)									
		2		3		4		5		6	
		F-Value	*p*-Value	*F*-Value	*p*-Value	*F*-Value	*p*-Value	*F*-Value	*p*-Value	*F*-Value	*p*-Value
ΔDBH (cm)	IF vs. CK	3.886	**0.006**	1.654	0.17	0.459	0.832	0.759	0.608	1.133	0.369
	I	15.88	**0.001**	0.387	0.061	0.009	0.926	0.063	0.804	0.106	0.748
	F	0.494	0.616	1.378	0.271	0.653	0.53	0.527	0.597	0.501	0.612
	I × F	0.463	0.635	0.147	0.864	0.021	0.979	1.278	0.297	1.848	0.179
ΔH (m)	IF vs. CK	1.247	0.313	1.177	0.347	0.727	0.632	1.230	0.321	0.865	0.533
	I	0.694	0.413	0.138	0.714	1.468	0.238	0.021	0.886	3.035	0.094
	F	0.746	0.485	2.32	0.12	0.898	0.421	3.711	**0.039**	0.343	0.713
	I × F	0.183	0.834	0.107	0.899	0.387	0.683	0.176	0.840	0.069	0.934
BA_S ($m^2 \cdot ha^{-1}$)	IF vs. CK	2.631	**0.038**	2.893	**0.025**	2.191	0.074	2.197	0.073	1.996	0.100
	I	8.628	**0.007**	9.877	**0.004**	6.445	**0.018**	9.227	**0.006**	8.814	**0.007**
	F	0.714	0.500	0.355	0.705	0.426	0.658	0.281	0.757	0.632	0.540
	I × F	0.633	0.539	0.678	0.517	0.449	0.644	1.548	0.233	0.178	0.838
V_S ($m^3 \cdot ha^{-1}$)	IF vs. CK	2.302	0.062	2.840	**0.027**	2.272	0.065	2.325	0.060	1.841	0.127
	I	6.229	**0.020**	9.662	**0.005**	6.878	**0.015**	9.207	**0.006**	7.894	**0.010**
	F	0.893	0.423	0.288	0.753	0.405	0.617	0.287	0.753	0.288	0.752
	I × F	0.472	0.629	0.608	0.553	0.516	0.603	1.522	0.239	0.241	0.788
AFP ($m^3 \cdot ha^{-1} \cdot yr^{-1}$)	IF vs. CK	2.302	0.062	2.840	**0.027**	2.272	0.065	2.325	0.060	1.841	0.127
	I	6.229	**0.020**	9.662	**0.005**	6.878	**0.015**	9.207	**0.006**	7.894	**0.010**
	F	0.893	0.423	0.288	0.753	0.405	0.617	0.287	0.753	0.288	0.752
	I × F	0.472	0.629	0.608	0.553	0.516	0.603	1.522	0.239	0.241	0.799

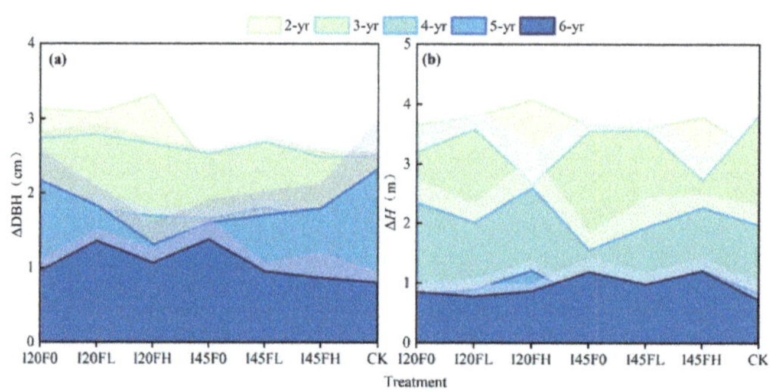

Figure 1. Effects of different treatments on the annual increment of diameter at breast height (**a**) and the annual increment of tree height (**b**) in 2–6-year-old *Populus tomentosa* stands. Note: The gray shaded area represents the standard error.

The BA_S of the 2–3-year-old water–nitrogen coupling treatments differed significantly from the CK treatment (Table 1). Except for the I45F0 treatment, the remaining five water–nitrogen treatments enhanced the stand's basal area. Compared to the CK treatment, the I20FH treatment significantly increased BA_S by 46.57% and 35.80% at 2 and 3 years of age, respectively ($p < 0.05$) (Figure 2). At 4–6 years old, no significant variations in BA_S

were found between treatments (Figure 2 and Table 1). The therapies had similar impacts on *Vs* and AFP as they did on BA_S. However, these significant differences did not show until 3 years old. Among the six water–fertilizer treatments, the BA_S, V_S, and AFP of the 2–6-year-old forest stand responded strongly to irrigation ($p < 0.05$) but were unaffected by fertilization or interaction effects.

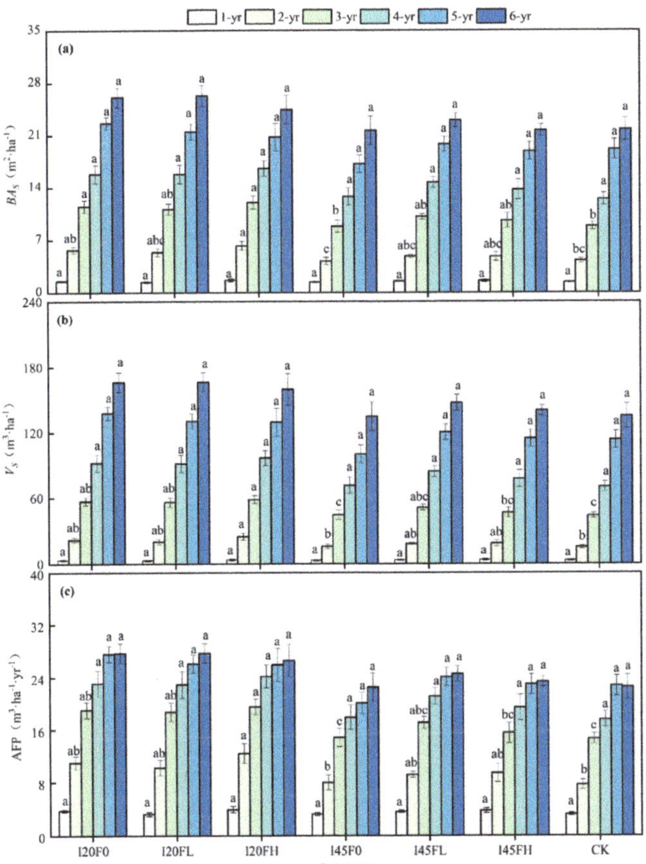

Figure 2. Effects of different treatments on stand basal area (**a**), stand volume (**b**), and average forest productivity (**c**) for 2–6-year-old *Populus tomentosa* stands. Note: Each value represents the mean of five replicates, with error bars indicating the standard error of the mean. A one-way ANOVA was used to test the effects of different treatments on stand growth indices within the same stand age, and multiple comparisons were conducted using a Duncan's test. Different lowercase letters indicate significant differences among treatments within the same year.

In the sixth year, there were no significant variations in BA_S, V_S, or AFP between the unfertilized treatments under different irrigation gradients (I20F0, I45F0, and CK). The same was true for the different fertilized treatments under high irrigation (I20F0, I20FL, and I20FH). However, for the low irrigation treatments (I45F0, I45FL, and I45FH), we found that *Vs* increased initially and then dropped as fertilization amount increased (Figure 3).

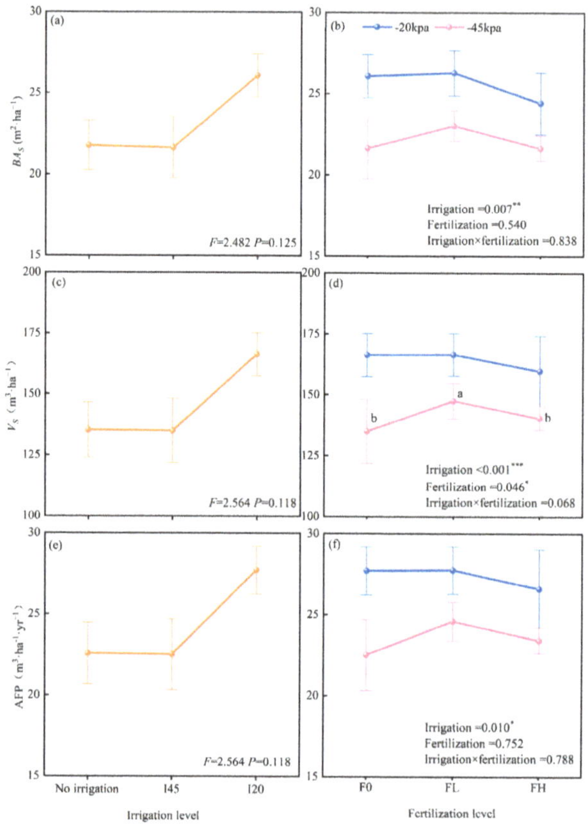

Figure 3. Effects of different irrigation gradients and fertilization levels on stand basal area (**a**,**b**), stand volume (**c**,**d**), and mean site productivity (**e**,**f**). Note: Each value represents the mean of five replicates, with error bars indicating the standard error of the mean. A two-way ANOVA was used to analyze the effects of irrigation level, fertilization level, and their interaction on various stand growth indicators. A one-way ANOVA was employed to test the effects of different irrigation levels and different fertilization levels under the same irrigation level on stand growth indicators, with multiple comparisons conducted using a Duncan's test. * indicates $p < 0.05$, ** indicates $p < 0.01$, *** indicates $p < 0.001$. Different lowercase letters indicate differences under various irrigation/fertilization levels.

2.2. Long-Term Effects of Different Water–Nitrogen Coupling Treatments on Biomass Distribution Pattern

In the sixth year, there was no significant difference in aboveground and belowground biomass allocation among treatments ($p > 0.05$). The above-ground biomass allocation exceeded the below-ground biomass allocation by 4.16 to 5.62 times (Figure 4a). The biomass allocation between trunks and branches varied among treatments, with full irrigation without fertilization reducing trunk biomass allocation (−5.37%) while boosting branch biomass allocation (+2.58%) (Figure 4b). Upper canopy branches (3.30~14.96%) had significantly reduced allocation compared to middle and lower canopy branches (34.51%~58.64%), with significant variations between treatments ($p < 0.05$) (Figure 5a). The low water irrigation (I45) and no irrigation (CK) treatments had considerably larger proportions of top canopy branches compared to the high water irrigation treatment (I20) ($p < 0.05$) (Figure 5a). Different root organs reacted differently to the treatments. Fine and coarse roots responded weakly to diverse treatments, with no significant differences between them ($p > 0.05$) (Figure 5b). The I20F0 treatment had a considerably larger allocation ratio for skeletal root biomass compared to the I20FH, I45FH, and CK treatments ($p < 0.05$) (Figure 5b).

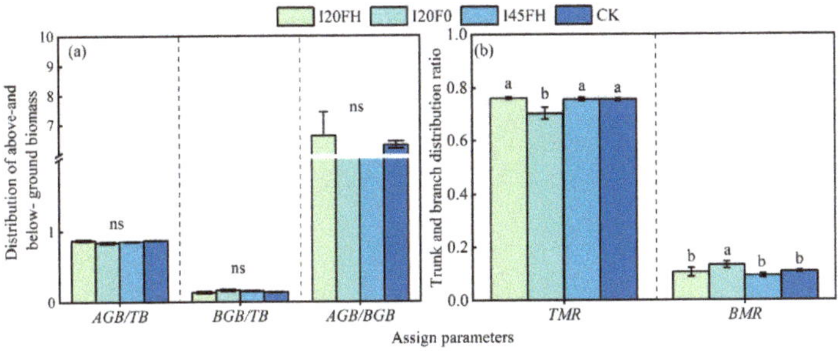

Figure 4. Effects of different treatments on the allocation ratio of above-ground to below-ground biomass (**a**) and the allocation ratio of biomass in stem and branch organs (**b**). Note: Each value represents the mean of three replicates, with error bars indicating the standard error of the mean. A one-way analysis of variance (ANOVA) was used to assess the effects of treatments on individual tree biomass allocation, followed by multiple comparisons using a Duncan's test. Different lowercase letters indicate significant differences between treatments.

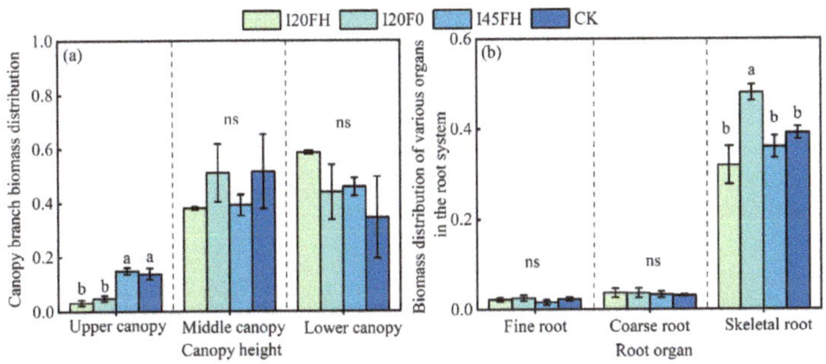

Figure 5. Effects of different treatments on the biomass allocation ratio of canopy branches (**a**) and the biomass allocation ratio of various root organs (**b**). Note: Each value represents the mean of three replicates, and the error bars indicate the standard error of the mean. A one-way ANOVA was used to test the effects of different treatments on individual tree biomass allocation, and multiple comparisons were conducted using a Duncan's test. Different lowercase letters indicate significant differences between treatments.

For the above-ground portion, the corresponding quadratic and linear regression models were $y = 0.00158x^2 + 0.04533x + 0.53249$ ($R^2 = 0.82$, $p < 0.001$) and $y = 0.01356x + 0.68006$ ($R^2 = 0.76$, $p < 0.001$). For the below-ground portion, the corresponding quadratic and linear regression models were $y = 0.00158x^2 - 0.04533x + 0.46751$ ($R^2 = 0.82$, $p < 0.001$) and $y = -0.01356x + 0.31994$ ($R^2 = 0.76$, $p < 0.001$). When the models' R^2 values are compared, the quadratic regression model surpasses the linear regression model and more closely reflects our observed data. Using the quadratic model, we discovered that when the *Populus tomentosa* diameter is small (younger in age), the allocation proportion to the aboveground part of the *Populus tomentosa* increases rapidly, while the below-ground allocation proportion declines. However, when the *Populus tomentosa*'s diameter rises, this proportion progressively stabilizes, and as the *Populus tomentosa* becomes larger, there is a pattern of diminishing aboveground allocation and increasing below-ground biomass allocation (Figure 6a,c).

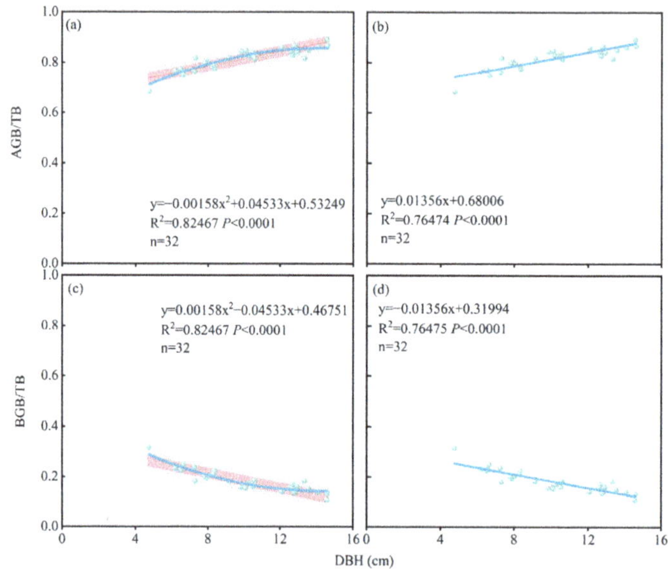

Figure 6. Linear and quadratic regression models of the proportion of aboveground biomass (**a**,**b**) and belowground biomass (**c**,**d**) of forest trees in relation to diameter at breast height (n = 32). Note: The pink shading represents the 95% prediction band, and the gray shading represents the 95% confidence band.

2.3. Long-Term Effects of Different Water–Nitrogen Coupling Treatments on WUE and PFP

In the sixth year, irrigation level had a significant effect on WUE, which was significantly lower in the I20 treatment than in the I45 treatment, with an average reduction of 88.79% under the three fertilization levels ($p < 0.001$) (Figure 7a). Fertilization with irrigation did not significantly promote WUE ($p > 0.05$) (Figure 7a), but there was a potential for WUE to initially increase and then decrease with increasing fertilization, which was more pronounced in the I45 treatment compared to the I20 treatment. Irrigation and fertilization levels had a substantial impact on PFP ($p < 0.05$) (Figure 7b). PFP showed a decreasing trend with the increase in fertilization level, with reductions of 130.95% and 132.86% for the I20 and I45 treatments, respectively. (Figure 7b).

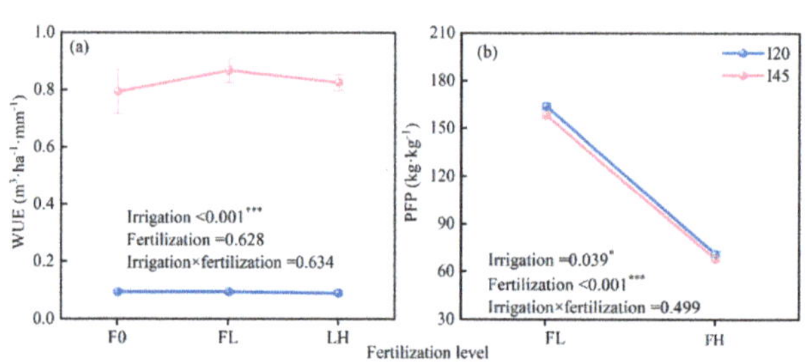

Figure 7. Effects of different irrigation gradients and fertilization levels on stand water use efficiency (WUE) (**a**) and partial factor productivity of nitrogen fertilizer (PFP) (**b**). Note: Each value represents the mean of five replicates, with error bars indicating the standard error of the mean. A two-way ANOVA was employed to analyze the effects of irrigation level, fertilization level, and their interaction on WUE and PFP. * indicates $p < 0.05$, *** indicates $p < 0.001$.

2.4. Correlation and Path Analysis

Correlation analysis showed that irrigation levels were considerably negatively connected with WUE ($p < 0.001$) (Figure 8a and Table S1) and strongly correlated with V_S, TGB, and AFP ($p < 0.01$) (Figure 8a and Table S1), with correlation values of 0.44, 0.42, and 0.44, respectively. Irrigation had a negative effect on WUE, with a direct path coefficient of -0.978 ($p < 0.01$) (Figure 8b), but had substantial beneficial effects on BA_S, V_S, AFP, and TGB, with direct path coefficients of 0.506, 0.489, 0.489, and 0.472, respectively. The fertilization level had a strong connection with PFP ($p < 0.01$) (Table S1), with a correlation coefficient of 0.99 (Figure 8a). The fertilization level had a negative impact on BA_S, V_S, AFP, TGB, WUE and PFP, and only a positive impact on H. Among the direct effects on various indicators, only PFP reached a significant level, with a direct path coefficient of 0.429 ($p < 0.01$) (Figure 8b). It is clear that irrigation exerts a more consistently beneficial influence on *Populus tomentosa* growth than fertilization.

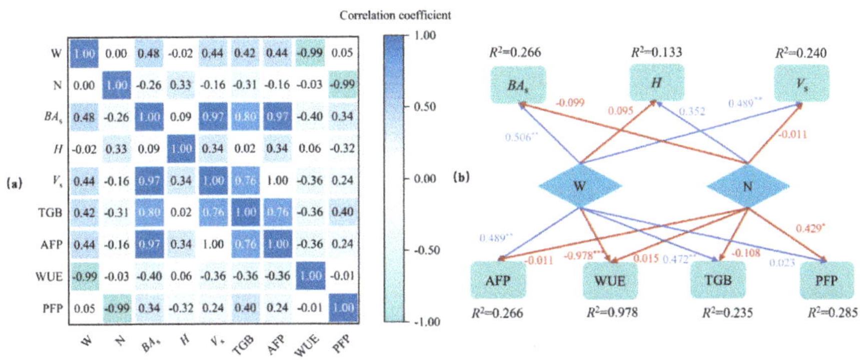

Figure 8. Correlation analysis (**a**) and path analysis (**b**) of irrigation (W) and fertilization (N) gradients with tree height (H), stand basal area (BA_S), stand volume (V_S) and productivity (AFP), forest biomass (TGB), water use efficiency (WUE), and partial factor productivity of nitrogen fertilizer (PFP). Note: Blue arrows represent positive effects; red arrows represent negative effects. *** indicates $p < 0.001$; ** indicates $p < 0.01$; * indicates $p < 0.05$.

3. Discussion

3.1. Long-Term Effects of Water–Nitrogen Coupling on Tree Growth

The productivity of forest land can be raised and tree growth accelerated by fertilizer and irrigation. However, most studies were often confined to specific stand ages, ignoring the different responses to fertilization and irrigation across various developmental stages within the forest rotation cycle [7–9,36–38]. The growth of short-rotation *Populus tomentosa* forest in this study showed an age-dependent response to various nitrogen and water treatments (Figures 1 and 2). Regarding *Populus tomentosa*, we discovered that the water–nitrogen coupling considerably increased the DBH development in trees that were two years old, but the response to the water–nitrogen coupling progressively ceased to be significant after three years of age (Figure 1 and Table 1). The variation in responses of different-aged *Populus tomentosa* to water–nitrogen coupling is primarily influenced by changes in tree living space. Irrigation and nitrogen application can hasten the development of the *Populus tomentosa*'s underground portions during the early phases of growth. But as the trees grow larger, the living space of forest trees increasingly limits the expansion of their underground portions, and their root systems gradually start to grow vertically along the soil profile. *Populus tomentosa* is a tree species with fine roots that can reach close to the groundwater level, according to earlier research [39,40]. The North China Plain is dominated by agricultural production, where fertilizer application frequently surpasses crop requirements. This results in elevated levels of NO_3^- and NH_4^+ ions in the groundwater when combined with the sandy soil texture [41]. To satisfy their own growth needs, trees draw water and nutrients from groundwater and deep soil layers [42,43]. Therefore, as the forest ages, *Populus tomentosa*'s ability to acquire deep soil resources through vertical root extension strengthens, significantly reducing its reliance on surface water and nitrogen inputs. Correspondingly, canopy competition effects become more pronounced with increasing stand density, as overlapping crowns mutually shade each other, impeding trees' capacity to absorb and transport soil moisture and inorganic salts due to diminished transpiration pull [2,44–46]. Another possible explanation for the weakening water–nitrogen coupling effect with increasing stand age stems from nutrient limitations caused by prolonged nitrogen input [47]. Specifically, sustained nitrogen addition may trigger secondary effects such as the depletion of soil base cations, enhanced aluminum ion activity, and soil acidification, which have been demonstrated to significantly inhibit forest growth [48,49].

This research also discovered that water–nitrogen coupling enhanced the effect of water inputs on stand growth compared to irrigation alone (Figure 3). Specifically, in the irrigation-only treatment, there were no significant differences in stand BA_S, V_S, and AFP with increasing irrigation amounts (Figure 3a,c,e). However, when comparing the six water–nitrogen treatments, we found that irrigation significantly improved the benefits of BA_S, V_S, and AFP, with particularly noticeable effects at lower fertilization rates (Figure 3b,d,f). According to stand characteristics at the conclusion of the rotation period, a lower fertilization treatment—rather than the maximum fertilization treatment—is the ideal fertilization rate under various irrigation gradients. More development is encouraged in the early stages by high fertilization (Figure 2). Long-term, though, this large fertilizer input results in decreased forest yield in addition to resource waste (Figure 3). Consequently, the forest age effect is the first element we cannot ignore in the intensive cultivation of water and nitrogen, followed by the trees' irrigation level. We should think about the nutrient variables and the water-to-fertilizer ratio only after the trees' water needs have been satisfied.

3.2. Long-Term Effects of Water–Nitrogen Coupling on Biomass Allocation

The total biomass ratio between the aboveground and belowground portions of the trees stayed comparatively constant throughout the various treatments as they grew and matured (Figure 4a). This supports the Isometric Growth Hypothesis by showing that the aboveground and belowground portions expanded at the same relative growth rate throughout the growth process [50–52]. The main limiting factor at the end of the rotation cycle is competition for living space, which progressively overtakes the availability of water and nutrients. According to the Functional Equilibrium Hypothesis and the Optimal Partitioning Hypothesis of biomass allocation, aboveground allocation gradually rises while belowground allocation falls (Figure 6a,c) [25,33,53]. In order to maximize their growth and reproductive benefits, plants typically optimize the distribution of biomass between aboveground and underground parts based on the availability of environmental resources like light, water, and nutrients [7,22,28,32,54–56]. Nevertheless, neither high water irrigation nor high fertilizer treatment upset the distribution ratio of the aboveground and below-ground biomass of forest trees, which would finally attain a balanced condition following long-term water and nitrogen coupling measures (Figure 4a). On the one hand, it might be because poplar growth in sandy soils is restricted by water and nitrogen, and the addition of these elements would encourage the development of the underground root system [57,58] (Figure 5b). Furthermore, in order to guarantee steady growth throughout the sapling period, fast-growing tree species need a more robust subterranean structure [39]. However, the accumulation of above-ground biomass is limited since the growing space for the forest trees' aboveground portion is gradually constricted.

We discovered some minor variations between the various organs, though. For example, plants lower biomass allocation to the higher canopy branches under high water treatment (I20) (Figure 5a). This is mainly because *Populus tomentosa* that receive enough irrigation grow more radially, which results in larger diameters, while their height growth is decreased [59]. Low irrigation treatments, on the other hand, encourage trees to grow vertically in order to compete for additional light resources, which results in more branches developing in the top canopy [60]. In the root system, we did not find that water and nitrogen treatments could regulate the biomass allocation of fine and coarse roots (Figure 5b). Nonetheless, the I20F0 treatment considerably increased the percentage of skeletal root biomass, whereas the high fertilizer treatment (FH) tended to decrease this percentage (Figure 5b). This also suggests that while long-term irrigation can improve the stability of tree root systems, adding nitrogen fertilizer carries the risk of reducing root stability and tree anchoring capacity. Likewise, we saw a definite relationship between the proportion of root systems to branches. The biomass allocation ratio of skeletal roots that serve as subsurface anchors increased in proportion to the aboveground branch allocation ratio (Figures 4b and 5b). Because larger branch systems are more likely to windthrow under windy conditions, the skeletal roots that serve as anchors also grow in an adaptable manner [61]. This further illustrates how functional matching and adaptation in plants are evolutionary processes [62].

3.3. Effects of Water–Nitrogen Coupling on WUE and PFP at the End of the Rotation Period

Water scarcity in China has been worse overall in recent years due to climate change, with the North China Plain experiencing an especially severe shortage [63,64]. Poplar plantations should have enough irrigation volume to maintain tree growth without wasting water resources [34,65]. One of the most pressing concerns in the irrigation management of poplar plantations in North China is the rational use of water resources and increasing water usage efficiency. This study's findings showed that when irrigation volume increases, the stand's WUE fell (Figures 7 and 8). Notably, the WUE of the stand was not significantly affected by varying nitrogen application gradients (Figures 7 and 8, Table S1). This result contrasts with that of Gu et al. [4], who found that increasing nitrogen treatment can improve winter rapeseed yield and WUE. This discovery may be connected to plant traits and developmental stages. According to earlier research [8,9], *Populus tomentosa* responded significantly to fertilizer and water during the seedling stage, allowing it to quickly acquire growth dominance. But as the forest matures, the trees became less sensitive to nitrogen and water, and the benefits of fertilizer and irrigation on growth and development declined as well (Table 1). Moreover, there was no significant difference in WUE under different fertilization treatments by the end of the rotation period.

PFP, in contrast to WUE, was controlled by fertilization and irrigation, although there was little interaction between them. This suggests that PFP is independently controlled by fertilization and irrigation rather than being impacted by the combined effects of water and nitrogen. At age six, irrigation could encourage PFP improvement, whereas more fertilization caused PFP to decline (Figure 7b). This is consistent with our results of tree growth at 5–6 years of age, indicating that PFP is not always enhanced by the maximal nitrogen supply to encourage wood accumulation. This also suggests that the sandy loam soil at this experimental site has a poor nutrient retention capacity, and that nitrogen fertilizers are prone to leaching, which results in a low partial productivity of nitrogen fertilizers and a lower rate of nitrogen fertilizer utilization by trees. In addition to posing hazards to ecological security, the likelihood of fertilizer leakage is anticipated to increase [66], which means that a high economic investment does not provide a similarly high output.

A reasonable stand density allows trees to make full use of resources including light, water, and nutrients [67]. Excessive stand density causes tree roots to intertwine and compress, resulting in competition for water and nutrients, reducing each tree's availability for these resources. Furthermore, the mutual shade of tree canopies reduces light availability for understory trees, lowering photosynthesis and decreasing the efficiency of water and nutrient utilization, affecting irrigation efficiency and nitrogen fertilizer partial production. As a result, stand density near the conclusion of the rotation period plays an important role in restricting water and nutrient efficiency, as well as tree growth. For semi-humid sandy short-rotation plantations, discontinuing irrigation and fertilizer at the end of the rotation period may be the most effective management method.

4. Materials and Methods

4.1. General Description of the Experimental Site

The experimental site is at the State-owned Jiucheng Forest Farm in in Gaotang County, Liaocheng City, China (36°48′47″ N, 116°05′25″ E) (Figure S1). *Populus tomentosa* is grown primarily in the North China Plain. The research area has a typical temperate continental semi-arid monsoon environment, with distinct seasons and plenty of sunlight [8]. The total annual sunshine duration is 2651.9 h, with 204 frost-free days. Rainfall is heaviest in July and August, accounting for 544.7 mm on average every year. The average annual temperature is 13.2 °C, with peaks of 41.2 °C and lows of −20.8 °C. The topography of the study site is dominated by plains, and the water table is 6 to 9 m deep [68]. The soil texture is sandy loam with 2.72% clay. Before afforestation, the average soil organic matter content in 0–80 cm was 0.334%, the total nitrogen content was 0.023%, and the available phosphorus content was 4.92 mg·kg^{-1}. For further details regarding the physical and chemical properties of the soil, refer to He et al. [8] and Liu et al. [68].

4.2. Experimental Design

This study focuses on the triploid Chinese white poplar S86 clone (*Populus tomentosa* × *Populus bolleana*) × (*Populus alba* × *Populus glandulosa*). In spring 2016, two-year-old bare-root stem cuttings were chosen for afforestation. The cuttings' diameter at breast height ranged from 2.12 to 3.51 cm, with an average of 2.68 cm, while their height ranged from 2.60 to 3.62 m, with an average of 3.30 m. The trees were planted equally at a spacing of 2 m × 3 m, yielding a planting density of 1666 trees per hectare. Each tree received 70 g of base fertilizer with N + P_2O_5 + K_2O ≥ 31% (14–12-5) and OM ≥ 15%. In 2016, the drip irrigation system was finished. The drip irrigation belts were placed 30 cm from the tree on both sides, in a two-line per row configuration. In 2017, the experiment with fertilization and irrigation began.

A completely randomized block design was used to establish the experimental layout with nitrogen application rate (F) and irrigation amount (I) as the treatment factors. This resulted in six water–nitrogen treatments (IF) and an additional natural control treatment (CK). The irrigation amount was set based on the quantitative relationship between the growth of *Populus tomentosa* and soil water availability [69]. The trees received irrigation when the soil water potential (SWP) at 20 cm below the dripper, as determined by a tensiometer, reached −20 kPa and −45 kPa (designated as I20 and I45, respectively). The rationale behind these irrigation thresholds was described in He et al. [8]. Based on the growth rhythm and nitrogen need characteristics of *Populus tomentosa*, fertilization rates were set [30]. Three nitrogen application levels were established: 0 (N0), low (NL: 80 kg ha^{-1} in 2017 and 120 kg ha^{-1} from 2018 onwards), and high (NH: 220 kg ha^{-1} in 2017 and 260 kg ha^{-1} from 2018 onwards). The fertilizer choice was urea with a 46% nitrogen content and N fertilizer—the urea nitrate solution was injected directly into a water-driven injector (MixRite Model 2504, Tefen, Israel). According to the biological characteristics of *Populus tomentosa*, which grows rapidly at first and then slows down, the first three applications accounted for 3/5 of the annual fertilization amount, while the last three applications accounted for 2/5 of the annual fertilization amount, with each application interval lasting approximately 20 days. Additionally, a control treatment (CK) without irrigation and fertilization was established. To avoid the influence of light and soil heterogeneity, each block was arranged longitudinally from north to south, with seven treatment plots per block, resulting in a total of five replicates and 35 plots. Each plot consisted of eight rows of trees, with four trees per row, and the middle four rows were designated as experimental trees, while the two outermost rows on the east and west sides served as protective rows. Furthermore, 50 cm deep plastic sheets were buried between adjacent plots within the same block to prevent the movement of water and nitrogen. Chemical agents were used regularly for weed and pest control. The amounts of irrigation and fertilization applied between 2017 and 2021 are shown in Table 2.

Table 2. Irrigation amount and fertilization amount under different treatments from 2017 to 2021.

Treatment	Irrigation When SWP at 20 cm Depth (kPa)	Irrigation Amount (mm·ha^{-1}·year^{-1})					Fertigation Amount (kg ha^{-1} year^{-1}N)				
		2017	2018	2019	2020	2021	2017	2018	2019	2020	2021
I20F0	−20	347.40	455.73	291.67	309.90	364.57	0	0	0	0	0
I20FL	−20	347.40	455.73	291.67	309.90	364.57	80	120	120	120	120
I20FH	−20	347.40	455.73	291.67	309.90	364.57	220	260	260	260	260
I45F0	−45	16.67	0.00	25.52	51.10	76.78	0	0	0	0	0
I45FL	−45	16.67	0.00	25.52	51.10	76.78	80	120	120	120	120
I45FH	−45	16.67	0.00	25.52	51.10	76.78	220	260	260	260	260
Control	Non-irrigation	0.00	0.00	0.00	0.00	0.00	0	0	0	0	0

Note: I20 and I45 indicate irrigation when the soil water potential reaches −20 kPa and −45 kPa, respectively, while F0, FL, and FH represent three levels of fertilization: no fertilization, low fertilization rate, and high fertilization. The irrigation volume represents the total amount of water applied under the irrigation scenario of the current year. Since the irrigation volume for fertigation is relatively small, it was not calculated in this study.

4.3. Growth Measurement and Biomass Acquisition

4.3.1. Acquisition of Forest Growth Data

From 2016 to 2021, during the growth cessation period after leaf fall (October), the diameter at breast height (DBH) of the sample trees in each experimental plot within the forest was measured using a diameter tape. These data were then used to determine the stand basal area (BA_S, m^2 ha^{-1}) at the end of each year. The height of the sample trees in each plot was measured using a laser hypsometer (Vertex IV, Haglöf, Länsele, Sweden). The annual increment in diameter at breast height (ΔDBH) and the annual increment in tree height (Δ*H*) were calculated using the following formulas:

$$\Delta \text{DBH} = \text{DBH2} - \text{DBH1} \quad (1)$$

$$\Delta H = H2 - H1 \quad (2)$$

Individual tree volume is calculated using the binary volume database for *Populus tomentosa* compiled by Chen [70]:

$$Va = 0.5134 \times H^{0.827} \times (\text{DBH}/100)^{1.9954} \quad (3)$$

DBH1 and DBH2 and $H1$ and $H2$ reflect the measured values of diameter at breast height and tree height over two years, respectively. Va represents the individual tree volume, H is the tree height, and DBH is the diameter at breast height.

The individual tree volume (Va) multiplied by the number of trees per hectare yields the stand volume per hectare (V_S). Then, dividing the stand volume per hectare by the number of planting years gives the annual forest productivity (AFP).

4.3.2. Acquisition and Calculation of Forest Biomass Data

In early March 2022, before the trees had sprouted and matured, three sample trees (with breast height diameters close to the plot average) were chosen from the I20FH, I20F0, I45FH, and CK plots for destructive biomass sampling. After the sample trees were felled, their total height and height to the first branch were measured, and they were classified into three strata based on crown length: upper, middle, and lower. The sample trees' root systems were manually excavated in a 2 m × 3 m area, centered on the stump, to a depth of 1 m, yielding 6 m^3 of excavated soil. Samples from each tree were classified as upper canopy branches, lower canopy branches, trunks, root stumps, skeleton roots (>5 mm), coarse roots (2–5 mm), and fine roots (<2 mm). Samples of each part were brought back to the laboratory and dried at 65 °C until the mass was unchanged. The moisture content of each part was calculated from the samples and converted to the dry weight of each organ. Aboveground biomass (AGB) is the combined biomass of branches and stems, whereas belowground biomass (BGB) is the combined biomass of rootstocks, skeletal roots, coarse roots, and fine roots. The total biomass of a plant (TB) is the sum of its constituent elements. The allocation ratio for each organ is computed using the biomass of the various organs, belowground biomass, and total plant biomass.

$$TMR = SB/TB \quad (4)$$

$$BMR = BB/TB \quad (5)$$

SB, *BB*, and *TB* represent the dry weight of biomass (kg) for the trunk, branches, and total plant, respectively.

In October 2017, four sample trees (DBH values comparable to the plot average) were chosen from the I20H and CK plots for destructive biomass sampling. In October 2018, three sample trees were chosen from the I20FH, I20F0, I45FH, and CK plots for destructive biomass sampling. A total of 32 sample trees were collected in 2017, 2018, and 2022. The models $y = ax + b$ and $y = ax^2 + bx + c$ were developed based on data on total plant biomass, aboveground and belowground biomass, and tree breast height diameter. Figure 7 and Figure S2 illustrate the findings of these models.

4.3.3. Irrigation Water Use Efficiency and Nitrogen Fertilizer Partial Productivity

The irrigation water use efficiency (WUE, $m^3 \cdot ha^{-1} \cdot mm^{-1}$) was determined using Equation (6). This study calculated the total irrigation water use efficiency since the initiation of irrigation in 2017. In this context, VS_{total} ($m^3 \cdot ha^{-1}$) represents the forest stand volume of the I20F0, I45F0, and CK treatments at the end of 2021, and *IA* (mm) is the total irrigation amount applied during the research period from 2017 to 2021.

$$WUE = VS_{total}/IA \tag{6}$$

We used R^2 as the metric for model goodness-of-fit, where a higher R^2 indicates better agreement between predicted and measured values [71,72]. The univariate quadratic regression and linear regression models for biomass–DBH yielded R^2 values of 0.9274 and 0.8950, respectively (Figure S2). Given the average DBH per treatment plot at the end of 2021, we selected the univariate quadratic regression model to calculate total biomass per plant under water–nitrogen coupling treatments. Equation (7) was used to evaluate the partial factor productivity of nitrogen fertilizer (PFP, kg·kg^{-1}). This study calculated the PFP since its application in 2017. TB_{total} (kg·ha^{-1}) is the total biomass per hectare determined by the model for I20FL, I20FH, I45FL, and I45FH, and T (kg·ha^{-}) is the total quantity of nitrogen fertilizer applied over the research period from 2017 to 2021.

$$PFP = TB_{total}/T \tag{7}$$

4.4. Statistics and Analysis of Data

The data were first organized using Excel 2019. A Levene's test was applied to verify the homogeneity of variances, and a Kolmogorov–Smirnov test was used to check for normality. The data were converted using the natural logarithm (ln) since the biomass allocation ratios, TMR, and AGB\RGB of the upper and intermediate canopy branches did not satisfy the requirements for an ANOVA. Data on biomass allocation and tree growth were subjected to a one-way ANOVA to examine the differences between water and nitrogen treatments and the control treatment. The effects of irrigation (I), nitrogen application (F), and their interaction (I × F) under water–nitrogen coupling (IF) treatment were examined using a two-way ANOVA on tree growth data, WUE, and PFP. Based on significant differences from CK, multiple comparisons were performed using a Duncan's test ($\alpha = 0.05$).

Using IBM SPSS Statistics software (SPSS25, IBM Inc., Chicago, IL, USA), a Spearman's correlation analysis and path analysis were performed using the irrigation level (W) and fertilization level (N) from treatments I20F0, I20FL, I20FH, I45F0, I45FL, and I45FH in 2021 as independent variables (X_j) and growth indicators, stock volume, forest productivity, forest biomass, water use efficiency, and nitrogen fertilizer partial productivity as dependent variables (Y_j). Standardized regression coefficients were interpreted as direct path coefficients (ci), since this study focuses only on the direct effects of X_j on Y_j. Origin software (Origin 2023, Origin Lab Corporation, Northampton, MA, USA) was used to plot the graphs.

5. Conclusions

This study showed that there is a considerable age effect in the effects of nitrogen and water treatments on short-rotation *Populus tomentosa* forestry development. The growth of DBH at the age of two years was greatly aided by the coupling effect of water and nitrogen at the individual level; the I20FH treatment dramatically increased DBH by 37.30%. However, beyond the age of three, this encouraging influence rapidly faded. *Populus tomentosa*'s aboveground–belowground biomass allocation ratio remains unchanged when water–nitrogen coupling is applied. The aboveground biomass allocation ranged from 5.16 to 6.62 times that of the belowground biomass. As the stand grows, the aboveground allocation increases and the belowground allocation diminishes toward the conclusion of the rotation period. Notably, the allocation ratio of branches to skeletal roots changes in a coordinated manner, indicating the evolutionary approach of plant functional adaptation. The irrigation level had a negative impact on the WUE of 6-year-old *Populus tomentosa*, with the I20 treatment reducing it by 88.79% compared to the I45 treatment. Additionally, due to the poor nutrient retention capacity of the sandy loam soil at the experimental site, nitrogen fertilizer is prone to leaching. As the fertilization rate increased, the PFP decreased by 130.95% to 132.86%.

Thus, it is recommended to implement full irrigation (irrigate when soil water potential reaches −20 kpa) and nitrogen fertilization for high-density, short-rotation *Populus tomentosa* plantations during the sapling stage (ages 1–3) in semi-humid sandy loam regions, while ceasing fertilization and irrigation management during the 4–6 year growth period. This strategy not only reduces resource waste, but it also maintains stand productivity. This work provides both a theoretical foundation and practical assistance for accurate water and fertilizer management in short-rotation plantations.

Supplementary Materials: The following supporting information can be downloaded at: https://www.mdpi.com/article/10.3390/plants14121833/s1, Table S1. Significance of correlations between irrigation and fertilization gradients and forest growth indicators and forestland indicators. In the table, W represents irrigation level, N represents fertilization level, BA_S represents tree basal area, H represents mean tree height, V_S represents stand volume in different years, TGB represents forestland biomass, AFP represents average forestland productivity, WUE represents water use efficiency of irrigation, PFP represents partial factor productivity of nitrogen fertilizer. *** indicates $p < 0.001$; ** indicates $p < 0.01$; * indicates $p < 0.05$. Figure S1. The positions of the experimental site and experimental plantations [colored areas in (a,b)]. The boundaries of our experimental plantation are indicated by red color (c). Figure S2. Univariate quadratic regression (a) and linear regression model (b) for total tree biomass and diameter at breast height. The pink shading represents the 95% prediction band, and the gray shading represents the 95% confidence band (n = 32).

Author Contributions: Formal analysis, Y.W.; Investigation, Y.W. and Y.H.; Data curation, Y.W.; Writing—original draft, B.X.; Writing—review & editing, Y.W., J.L., Y.H., W.Z. and L.J.; Visualization, Y.W. and J.L.; Supervision, J.L.; Funding acquisition, L.J. and B.X. All authors have read and agreed to the published version of the manuscript.

Funding: This research was funded by the Special Foundation for National Science and Technology Basic Research Program of China (No. 2021YFD220120302) and the National Natural Science Foundation of China (No. 3167030059).

Data Availability Statement: The original contributions presented in this study are included in the article/Supplementary Materials. Further inquiries can be directed to the corresponding author(s).

Acknowledgments: We thank the original contributions put forward in this study, which have been incorporated into the article's materials. Further inquiries can be directed to the corresponding author/s. L.J. and B.X. originated the research concept and crafted the experimental framework. We express our heartfelt gratitude to Jinge Sun for his tending of the plants.

Conflicts of Interest: This research was conducted without any commercial or financial relationships that could be construed as potential conflicts of interest.

References

1. Warman, R.D. Global Wood Production from Natural Forests Has Peaked. *Biodivers. Conserv.* **2014**, *23*, 1063–1078. [CrossRef]
2. Forrester, D.I. Growth Responses to Thinning, Pruning and Fertiliser Application in Eucalyptus Plantations: A Review of Their Production Ecology and Interactions. *For. Ecol. Manag.* **2013**, *310*, 336–347. [CrossRef]
3. Zhu, W.; Wang, Y.; Qu, G.; Xi, B.; Jia, L. Cultivar and Tree Size, but Not Climate, Are Principal Factors Affecting Stem Quality of Populus Tomentosa Plantations in the North China Plain. *For. Ecol. Manag.* **2024**, *573*, 122348. [CrossRef]

4. Gu, X.-B.; Li, Y.-N.; Du, Y.-D. Effects of Ridge-Furrow Film Mulching and Nitrogen Fertilization on Growth, Seed Yield and Water Productivity of Winter Oilseed Rape (*Brassica napus* L.) in Northwestern China. *Agric. Water Manag.* **2018**, *200*, 60–70. [CrossRef]
5. Ahmed, A.K.M.; Jiang, L.; Wang, F.; Chen, S.; Zhou, X.; Pei, X.; Zhao, X.; Qu, G. Variation Analysis of Growth Traits of Four Poplar Clones under Different Water and Fertilizer Management. *J. For. Res.* **2020**, *31*, 45–55. [CrossRef]
6. Fu, Y.; Ding, G.; Quan, W.; Zhao, X.; Tao, Q. Coupling Optimization of Water-Fertilizer for Coordinated Development of the Environment and Growth of Pinus Massoniana Seedlings. *Agric. Water Manag.* **2024**, *300*, 108895. [CrossRef]
7. Coyle, D.R.; Coleman, M.D. Forest Production Responses to Irrigation and Fertilization Are Not Explained by Shifts in Allocation. *For. Ecol. Manag.* **2005**, *208*, 137–152. [CrossRef]
8. He, Y.; Xi, B.; Li, G.; Wang, Y.; Jia, L.; Zhao, D. Influence of Drip Irrigation, Nitrogen Fertigation, and Precipitation on Soil Water and Nitrogen Distribution, Tree Seasonal Growth and Nitrogen Uptake in Young Triploid Poplar (Populus Tomentosa) Plantations. *Agric. Water Manag.* **2021**, *243*, 106460. [CrossRef]
9. He, Y.; Xi, B.; Bloomberg, M.; Jia, L.; Zhao, D. Effects of Drip Irrigation and Nitrogen Fertigation on Stand Growth and Biomass Allocation in Young Triploid Populus Tomentosa Plantations. *For. Ecol. Manag.* **2020**, *461*, 117937. [CrossRef]
10. Van Den Driessche, R.; Thomas, B.R.; Kamelchuk, D.P. Effects of N, NP, and NPKS Fertilizers Applied to Four-Year Old Hybrid Poplar Plantations. *New For.* **2008**, *35*, 221–233. [CrossRef]
11. Rappe-George, M.O.; Gärdenäs, A.I.; Kleja, D.B. The Impact of Four Decades of Annual Nitrogen Addition on Dissolved Organic Matter in a Boreal Forest Soil. *Biogeosci. Discuss.* **2012**, *9*, 12433. [CrossRef]
12. Jiang, L.; Tian, D.; Ma, S.; Zhou, X.; Xu, L.; Zhu, J.; Jing, X.; Zheng, C.; Shen, H.; Zhou, Z.; et al. The Response of Tree Growth to Nitrogen and Phosphorus Additions in a Tropical Montane Rainforest. *Sci. Total Environ.* **2018**, *618*, 1064–1070. [CrossRef] [PubMed]
13. Goswami, S.; Fisk, M.C.; Vadeboncoeur, M.A.; Garrison-Johnston, M.; Yanai, R.D.; Fahey, T.J. Phosphorus Limitation of Aboveground Production in Northern Hardwood Forests. *Ecology* **2018**, *99*, 438–449. [CrossRef] [PubMed]
14. DeForest, J.L.; Snell, R.S. Tree Growth Response to Shifting Soil Nutrient Economy Depends on Mycorrhizal Associations. *New Phytol.* **2020**, *225*, 2557–2566. [CrossRef]
15. Nasto, M.K.; Winter, K.; Turner, B.L.; Cleveland, C.C. Nutrient Acquisition Strategies Augment Growth in Tropical N_2 -fixing Trees in Nutrient-poor Soil and under Elevated CO_2. *Ecology* **2019**, *100*, e02646. [CrossRef]
16. Liu, W.; Duan, H.; Shen, F.; Liao, Y.; Li, Q.; Wu, J. Effects of Long-term Nitrogen Addition on Water Use by *Cunninghamia Lanceolate* in a Subtropical Plantation. *Ecosphere* **2022**, *13*, e4033. [CrossRef]
17. Zhang, W.; Lu, J.-S.; Bai, J.; Khan, A.; Zhao, L.; Wang, W.; Zhu, S.-G.; Liu, S.-T.; Jin, J.-M.; Nyanchera, G.D.; et al. Reduced Fertilization Boosts Soil Quality and Economic Benefits in Semiarid Apple Orchard: A Two-Year Appraisal of Fertigation Strategy. *Agric. Water Manag.* **2024**, *295*, 108766. [CrossRef]
18. Zhang, R.; Nie, L.; Huang, M.; Yang, H.; Shi, C.; Wei, Y.; Song, L.; Zhu, J.; Bo, H.; Wang, J.; et al. Effects of Irrigation and Nitrogen Application on Soil Nutrients in Triploid Populus Tomentosa Stands. *Forests* **2022**, *13*, 1046. [CrossRef]
19. Roa-Fuentes, L.L.; Campo, J.; Parra-Tabla, V. Plant Biomass Allocation across a Precipitation Gradient: An Approach to Seasonally Dry Tropical Forest at Yucatán, Mexico. *Ecosystems* **2012**, *15*, 1234–1244. [CrossRef]
20. Wu, J.; Hong, J.; Wang, X.; Sun, J.; Lu, X.; Fan, J.; Cai, Y. Biomass Partitioning and Its Relationship with the Environmental Factors at the Alpine Steppe in Northern Tibet. *PLoS ONE* **2013**, *8*, e81986. [CrossRef]
21. Bebre, I.; Marques, I.; Annighöfer, P. Biomass Allocation and Leaf Morphology of Saplings Grown under Various Conditions of Light Availability and Competition Types. *Plants* **2022**, *11*, 305. [CrossRef] [PubMed]
22. Kara, F. Effects of Light Transmittance on Growth and Biomass of Understory Seedlings in Mixed Pine-Beech Forests. *Eur. J. For. Res.* **2022**, *141*, 1189–1200. [CrossRef]
23. Qi, Y.; Wei, W.; Chen, C.; Chen, L. Plant Root-Shoot Biomass Allocation over Diverse Biomes: A Global Synthesis. *Glob. Ecol. Conserv.* **2019**, *18*, e00606. [CrossRef]
24. Zhang, H.; Song, T.; Wang, K.; Yang, H.; Yue, Y.; Zeng, Z.; Peng, W.; Zeng, F. Influences of Stand Characteristics and Environmental Factors on Forest Biomass and Root–Shoot Allocation in Southwest China. *Ecol. Eng.* **2016**, *91*, 7–15. [CrossRef]
25. Poorter, H.; Jagodzinski, A.M.; Ruiz-Peinado, R.; Kuyah, S.; Luo, Y.; Oleksyn, J.; Usoltsev, V.A.; Buckley, T.N.; Reich, P.B.; Sack, L. How Does Biomass Distribution Change with Size and Differ among Species? An Analysis for 1200 Plant Species from Five Continents. *New Phytol.* **2015**, *208*, 736–749. [CrossRef]
26. Baret, M.; Pepin, S.; Ward, C.; Pothier, D. Long-Term Changes in Belowground and Aboveground Resource Allocation of Boreal Forest Stands. *For. Ecol. Manag.* **2015**, *350*, 62–69. [CrossRef]
27. Byambadorj, S.-O.; Nyam-Osor, B.; Park, B.B.; Avirmed, T.; Scippa, G.S.; Chiatante, D.; Montagnoli, A.; Dimitrova, A. Afforestation of Mongolian Steppe: Patterns of Biomass Partitioning in *Populus Sibirica* and *Ulmus Pumila* Trees in Response to Management Supporting Measures. *Plant Biosyst.-Int. J. Deal. All. Asp. Plant Biol.* **2022**, *156*, 969–981. [CrossRef]

28. Fang, S.; Xu, X.; Lu, S.; Tang, L. Growth Dynamics and Biomass Production in Short-Rotation Poplar Plantations: 6-Year Result for Three Clones at Four Spacings. *Biomass Bioenergy* **1999**, *17*, 415–425. [CrossRef]
29. Lie, Z.; Xue, L.; Jacobs, D.F. Allocation of Forest Biomass across Broad Precipitation Gradients in China's Forests. *Sci. Rep.* **2018**, 10536. [CrossRef]
30. Wang, Y.; Xi, B.; Bloomberg, M.; Moltchanova, E.; Li, G.; Jia, L. Response of Diameter Growth, Biomass Allocation and N Uptak to N Fertigation in a Triploid Populus Tomentosa Plantation in the North China Plain: Ontogenetic Shift Does Not Exclud Plasticity. *Eur. J. For. Res.* **2015**, *134*, 889–898. [CrossRef]
31. Wang, Q.-W.; Daumal, M.; Nagano, S.; Yoshida, N.; Morinaga, S.-I.; Hikosaka, K. Plasticity of Functional Traits and Optimality o Biomass Allocation in Elevational Ecotypes of Arabidopsis Halleri Grown at Different Soil Nutrient Availabilities. *J. Plant Res* **2019**, *132*, 237–249. [CrossRef]
32. Oliveira, N.; Rodríguez-Soalleiro, R.; Pérez-Cruzado, C.; Cañellas, I.; Sixto, H.; Ceulemans, R. Above- and below-Ground Carbo Accumulation and Biomass Allocation in Poplar Short Rotation Plantations under Mediterranean Conditions. *For. Ecol. Manag* **2018**, *428*, 57–65. [CrossRef]
33. Reich, P.B.; Luo, Y.; Bradford, J.B.; Poorter, H.; Perry, C.H.; Oleksyn, J. Temperature Drives Global Patterns in Forest Biomas Distribution in Leaves, Stems, and Roots. *Proc. Natl. Acad. Sci. USA* **2014**, *111*, 13721–13726. [CrossRef] [PubMed]
34. Xi, B.; Clothier, B.; Coleman, M.; Duan, J.; Hu, W.; Li, D.; Di, N.; Liu, Y.; Fu, J.; Li, J.; et al. Irrigation Management in Popla (Populus Spp.) Plantations: A Review. *For. Ecol. Manag.* **2021**, *494*, 119330. [CrossRef]
35. Wang, M.; Li, G.; Feng, Z.; Liu, Y.; Xu, Y.; Uscola, M. Uptake of Nitrogen Forms by Diploid and Triploid White Poplar Depends o Seasonal Carbon Use Strategy and Elevated Summer Ozone. *J. Exp. Bot.* **2021**, *72*, 7180–7190. [CrossRef]
36. Coleman, M. Spatial and Temporal Patterns of Root Distribution in Developing Stands of Four Woody Crop Species Grown wit Drip Irrigation and Fertilization. *Plant Soil.* **2007**, *299*, 195–213. [CrossRef]
37. Yan, X.-L.; Dai, T.-F.; Jia, L.-M. Evaluation of the Cumulative Effect of Drip Irrigation and Fertigation on Productivity in a Popla Plantation. *Ann. For. Sci.* **2018**, *75*, 5. [CrossRef]
38. Campoe, O.C.; Stape, J.L.; Albaugh, T.J.; Lee Allen, H.; Fox, T.R.; Rubilar, R.; Binkley, D. Fertilization and Irrigation Effects on Tre Level Aboveground Net Primary Production, Light Interception and Light Use Efficiency in a Loblolly Pine Plantation. *For. Eco Manag.* **2013**, *288*, 43–48. [CrossRef]
39. Zou, S.; Li, D.; Di, N.; Liu, J.; Li, L.; Liu, Y.; Xi, B.; Coleman, M. Stand Development Modifies Effects of Soil Water Availability o Poplar Fine-Root Traits: Evidence from a Six-Year Experiment. *Plant Soil.* **2022**, *480*, 165–184. [CrossRef]
40. Tan, J.; Yu, W.; Liu, Y.; Guo, Y.; Liu, N.; Fu, H.; Di, N.; Duan, J.; Li, X.; Xi, B. Response of Fine-Root Traits of Populus Tomentosa t Drought in Shallow and Deep Soil. *Forests* **2023**, *14*, 951. [CrossRef]
41. Wang, S.; Zheng, W.; Currell, M.; Yang, Y.; Zhao, H.; Lv, M. Relationship between Land-Use and Sources and Fate of Nitrate i Groundwater in a Typical Recharge Area of the North China Plain. *Sci. Total Environ.* **2017**, *609*, 607–620. [CrossRef] [PubMed]
42. Pinheiro, R.C.; Bouillet, J.-P.; Regina Pivello, V.; Aló, L.L.; Costa, V.E.; Van Den Meersche, K.; Guerrini, I.A.; Laclau, J.-P. Root Take up Labeled Nitrogen from a Depth of 9 m in a Wooded Savanna in Brazil. *Soil. Biol. Biochem.* **2021**, *160*, 108282. [CrossRef]
43. Henriksson, N.; Lim, H.; Marshall, J.; Franklin, O.; McMurtrie, R.E.; Lutter, R.; Magh, R.; Lundmark, T.; Näsholm, T. Tree Wate Uptake Enhances Nitrogen Acquisition in a Fertilized Boreal Forest—But Not under Nitrogen-poor Conditions. *New Phytol.* **2021** *232*, 113–122. [CrossRef]
44. Forrester, D.I.; Collopy, J.J.; Beadle, C.L.; Warren, C.R.; Baker, T.G. Effect of Thinning, Pruning and Nitrogen Fertiliser Applicatio on Transpiration, Photosynthesis and Water-Use Efficiency in a Young Eucalyptus Nitens Plantation. *For. Ecol. Manag.* **2012**, *266* 286–300. [CrossRef]
45. Forrester, D.I.; Collopy, J.J.; Beadle, C.L.; Baker, T.G. Effect of Thinning, Pruning and Nitrogen Fertiliser Application on Light Interception and Light-Use Efficiency in a Young Eucalyptus Nitens Plantation. *For. Ecol. Manag.* **2013**, *288*, 21–30. [CrossRef]
46. Brinkhoff, R.; Mendham, D.; Hunt, M.; Unwin, G.; Hovenden, M. Nitrogen Fertiliser Only Increases Leaf Area in the Lower Crown of Mid-Rotation Eucalyptus Nitens Plantations. *For. Ecol. Manag.* **2022**, *508*, 120048. [CrossRef]
47. Lu, X.; Vitousek, P.M.; Mao, Q.; Gilliam, F.S.; Luo, Y.; Zhou, G.; Zou, X.; Bai, E.; Scanlon, T.M.; Hou, E.; et al. Plant Acclimation to Long-Term High Nitrogen Deposition in an N-Rich Tropical Forest. *Proc. Natl. Acad. Sci. USA* **2018**, *115*, 5187–5192. [CrossRef]
48. Mao, Q.; Lu, X.; Zhou, K.; Chen, H.; Zhu, X.; Mori, T.; Mo, J. Effects of Long-Term Nitrogen and Phosphorus Additions on Soil Acidification in an N-Rich Tropical Forest. *Geoderma* **2017**, *285*, 57–63. [CrossRef]
49. Lu, X.; Mo, J.; Gilliam, F.S.; Zhou, G.; Fang, Y. Effects of Experimental Nitrogen Additions on Plant Diversity in an Old-growth Tropical Forest. *Glob. Change Biol.* **2010**, *16*, 2688–2700. [CrossRef]
50. Enquist, B.J.; Niklas, K.J. Global Allocation Rules for Patterns of Biomass Partitioning in Seed Plants. *Science* **2002**, *295*, 1517–1520 [CrossRef]

51. Yue, K.; Fornara, D.A.; Li, W.; Ni, X.; Peng, Y.; Liao, S.; Tan, S.; Wang, D.; Wu, F.; Yang, Y. Nitrogen Addition Affects Plant Biomass Allocation but Not Allometric Relationships among Different Organs across the Globe. *J. Plant Ecol.* **2021**, *14*, 361–371. [CrossRef]
52. Niklas, K.J. Modelling Below- and Above-Ground Biomass for Non-Woody and Woody Plants. *Ann. Bot.* **2005**, *95*, 315–321. [CrossRef]
53. Mccarthy, M.C.; Enquist, B.J. Consistency between an Allometric Approach and Optimal Partitioning Theory in Global Patterns of Plant Biomass Allocation. *Funct. Ecol.* **2007**, *21*, 713–720. [CrossRef]
54. Zhou, L.; Hong, Y.; Li, C.; Lu, C.; He, Y.; Shao, J.; Sun, X.; Wang, C.; Liu, R.; Liu, H.; et al. Responses of Biomass Allocation to Multi-Factor Global Change: A Global Synthesis. *Agric. Ecosyst. Environ.* **2020**, *304*, 107115. [CrossRef]
55. He, L.; Zhang, X.; Wang, X.; Ullah, H.; Liu, Y.; Duan, J. Tree Crown Affects Biomass Allocation and Its Response to Site Conditions and the Density of Platycladus Orientalis Linnaeus Plantation. *Forests* **2023**, *14*, 2433. [CrossRef]
56. Yang, Y.; Fang, J.; Ji, C.; Han, W. Above- and Belowground Biomass Allocation in Tibetan Grasslands. *J. Veg. Sci.* **2009**, *20*, 177–184. [CrossRef]
57. He, Y.; Li, G.; Xi, B.; Zhao, H.; Jia, L. Fine Root Plasticity of Young Populus Tomentosa Plantations under Drip Irrigation and Nitrogen Fertigation in the North China Plain. *Agric. Water Manag.* **2022**, *261*, 107341. [CrossRef]
58. Jia, S.; Meng, F.; Li, X. Soil Nutrients and Season Drive Fine Root Traits, Soil Microbial Community and Their Interaction in a Pinus Koraiensis Seed Orchard. *For. Ecol. Manag.* **2025**, *580*, 122503. [CrossRef]
59. Zhu, W.; Wu, X.; Jia, L.; Xi, B. Effects of Key Forest Management Practices and Climatic Factors on the Growth of Populus Tomentosa Plantations in the North China Plain. *For. Ecol. Manag.* **2022**, *521*, 120444. [CrossRef]
60. Kröner, K.; Larysch, E.; Schindler, Z.; Obladen, N.; Frey, J.; Stangler, D.F.; Seifert, T. Influence of Crown Morphology and Branch Architecture on Tree Radial Growth of Drought-Affected *Fagus sylvatica* L. *For. Ecosyst.* **2024**, *11*, 100237. [CrossRef]
61. Samariks, V.; Brizga, D.; Rūba, J.; Seipulis, A.; Jansons, Ā. Root-Plate Characteristics of Common Aspen in Hemiboreal Forests of Latvia: A Case Study. *Forests* **2020**, *12*, 32. [CrossRef]
62. Hassan, S.; Liguori, G.; Inglese, P.; Louhaichi, M.; Sortino, G. The Effect of Soil Volume Availability on Opuntia Ficus-Indica Canopy and Root Growth. *Agronomy* **2020**, *10*, 635. [CrossRef]
63. Piao, S.; Ciais, P.; Huang, Y.; Shen, Z.; Peng, S.; Li, J.; Zhou, L.; Liu, H.; Ma, Y.; Ding, Y.; et al. The Impacts of Climate Change on Water Resources and Agriculture in China. *Nature* **2010**, *467*, 43–51. [CrossRef]
64. Haddeland, I.; Heinke, J.; Biemans, H.; Eisner, S.; Flörke, M.; Hanasaki, N.; Konzmann, M.; Ludwig, F.; Masaki, Y.; Schewe, J.; et al. Global Water Resources Affected by Human Interventions and Climate Change. *Proc. Natl. Acad. Sci. USA* **2014**, *111*, 3251–3256. [CrossRef]
65. Xi, B.; Di, N.; Wang, Y.; Duan, J.; Jia, L. Modeling Stand Water Use Response to Soil Water Availability and Groundwater Level for a Mature Populus Tomentosa Plantation Located on the North China Plain. *For. Ecol. Manag.* **2017**, *391*, 63–74. [CrossRef]
66. Gu, B.; Ge, Y.; Chang, S.X.; Luo, W.; Chang, J. Nitrate in Groundwater of China: Sources and Driving Forces. *Glob. Environ. Change* **2013**, *23*, 1112–1121. [CrossRef]
67. Forrester, D.I.; Collopy, J.J.; Beadle, C.L.; Baker, T.G. Interactive Effects of Simultaneously Applied Thinning, Pruning and Fertiliser Application Treatments on Growth, Biomass Production and Crown Architecture in a Young Eucalyptus Nitens Plantation. *For. Ecol. Manag.* **2012**, *267*, 104–116. [CrossRef]
68. Liu, J.; Li, D.; Fernández, J.-E.; Coleman, M.; Hu, W.; Di, N.; Zou, S.; Liu, Y.; Xi, B.; Clothier, B. Variations in Water-Balance Components and Carbon Stocks in Poplar Plantations with Differing Water Inputs over a Whole Rotation: Implications for Sustainable Forest Management under Climate Change. *Agric. For. Meteorol.* **2022**, *320*, 108958. [CrossRef]
69. Xi, B.; Bloomberg, M.; Watt, M.S.; Wang, Y.; Jia, L. Modeling Growth Response to Soil Water Availability Simulated by HYDRUS for a Mature Triploid Populus Tomentosa Plantation Located on the North China Plain. *Agric. Water Manag.* **2016**, *176*, 243–254. [CrossRef]
70. Chen, Z. Compilation of Poplar Two-dimensional Standing Wood Volume table. *For. Res.* **1989**, *01*, 78–83. [CrossRef]
71. Liu, J.; Wang, D.; Yan, X.; Jia, L.; Chen, N.; Liu, J.; Zhao, P.; Zhou, L.; Cao, Q. Effect of Nitrogen, Phosphorus and Potassium Fertilization Management on Soil Properties and Leaf Traits and Yield of Sapindus Mukorossi. *Front. Plant Sci.* **2024**, *15*, 1300683. [CrossRef] [PubMed]
72. Xia, P.; Guo, H.; Zhao, H.; Jiao, J.; Deyholos, M.K.; Yan, X.; Liu, Y.; Liang, Z. Optimal Fertilizer Application for Panax Notoginseng and Effect of Soil Water on Root Rot Disease and Saponin Contents. *J. Ginseng. Res.* **2016**, *40*, 38–46. [CrossRef] [PubMed]

plants

MDPI

Article

Optimization of Irrigation Amount and Nitrogen Rate of Drip-Fertigated Sugar Beet Based on Sugar Yield, Nitrogen Use Efficiency, and Critical Nitrogen Dilution Curve in the Arid Southern Xinjiang of China

Ying Wang, Fulai Yan *, Junliang Fan and Fucang Zhang *

Key Laboratory of Agricultural Soil and Water Engineering in Arid and Semiarid Areas of Ministry of Education, Northwest A&F University, Yangling 712100, China; wangying930313@163.com (Y.W.); nwwfjl@163.com (J.F.)
* Correspondence: fulai_yan@163.com (F.Y.); zhangfc@nwsuaf.edu.cn (F.Z.)

Abstract: The critical nitrogen (N) dilution curve is widely used to diagnose crop N status, but no such model has been developed for sugar beet. This study evaluated the effects of irrigation amount and N rate on sugar yield, N use efficiency, and soil nitrate-N (NO_3-N) residue of drip-fertigated sugar beet in the arid southern Xinjiang of China. A reliable N nutrition index (NNI) for sugar yield was also established based on a critical N dilution curve derived from the dry matter of sugar beet. A three-year field experiment was established with six N rates (25–480 kg N ha^{-1}) and three irrigation levels based on crop evapotranspiration (ET_c) (0.6, 0.8, and 1.0 ET_c in 2019 and 2020, and 0.4, 0.6, and 0.8 ET_c in 2021). Results showed that sugar yield and N uptake increased and then generally stabilized with increasing N rate, while N use efficiency decreased. Most soil NO_3-N was mainly distributed in the 0–60 cm soil layer, but increasing irrigation amount reduced residual NO_3-N in the 0–80 cm soil layer. Additionally, the established critical N dilution curve of sugar beet was considered stable (Normalized RMSE = 16.6%), and can be used to calculate plant N requirements and further N rates during sugar beet growth. The results indicated that the optimal NNI was 0.97 under 0.6 ET_c for sugar yield production of sugar beet in this study. This study provides a basis for efficient water and N management in sugar beet production in arid and semi-arid regions globally.

Keywords: sugar yield; nitrogen (N) use efficiency; N concentration; N nutrition index; soil nitrate nitrogen

1. Introduction

Irrigation and nitrogen (N) application are the two most important agricultural practices for improving crop growth and yield in arid and semi-arid areas [1–3]. In modern agricultural production, it is common that people apply excessive irrigation and N fertilizer to maximize crop yields [4,5]. Recently, fertigation using partially treated wastewater has also been explored as a way to recycle water and nutrients in arid regions [6]. However, these agricultural practices ignore the synchronization between plant N demand and supply [7,8]. Many researchers have reported that this practice not only reduces crop yield and N use efficiency but also deteriorates the soil environment and poses a serious threat to the sustainable development of agriculture [9–11]. However, insufficient irrigation and N supply also limit plant growth and decrease crop yields [12]. Therefore, it is crucial to determine the required amount of irrigation and N fertilization based on the estimated water and N requirements of a specific crop [13,14].

Sugar beet (*Beta vulgaris* L.) is a common industrial crop that is salt-tolerant [15] and has a high economic benefit for farmers. Drip irrigation under plastic mulch is widely used for agricultural production in arid and semi-arid regions because this technology can alleviate the shortage of agricultural water resources and stimulate the sustainable production of local crops. For this reason, sugar beet has been widely cultivated in Xinjiang, Northwest China. Sugar beet has high water requirements, and the sugar yield of sugar beet is greatly affected by the N rate [16,17]. Thus, ensuring a balanced supply of water and N fertilizer can lead to high sugar yields [18,19]. However, people in the regions blindly apply irrigation water and N fertilizer, resulting in stunted growth of sugar beet [20], a decline in yield, water and fertilizer use efficiencies, and a waste of water and fertilizer resources [12,21], with the attendant environmental problems. The amount of N applied to sugar beet depends on factors such as the demand for N in the forms of nitrate (NO_3^-) and ammonium (NH_4^+) for crop growth, as well as the availability of these N forms in the soil [7,22], which indicates that the optimal N rate required in sugar beet production is uncertain [8]. Early on, Bu et al. [23,24] used remote sensing technology to monitor N concentration as a guide to predict the yield of sugar beet. However, it is difficult to apply this technology in most areas due to its high cost.

In previous studies, there has been a lot of focus on the accurate diagnosis of plant N concentration, including the critical N (N_c) dilution curve of crops and the N nutrition index (NNI) [25–27]. The N_c dilution curve and NNI are usually used to evaluate crop N requirements for reliable N recommendations [28]. The N_c dilution curve can be defined as the minimum N concentration at the maximum biomass during a certain growth period, which is usually based on the dry matter of stems and leaves or the whole plant [29]. Calculating the NNI can quantify crop N status, which is a reliable indicator for evaluating crop N nutrition status [30]. Previous studies have reported different N_c dilution curve models, including general ones and specific ones for C3 or C4 crops [31]. However, there are differences in the N uptake dynamics of different crops, and there is also an effect of climate. When a model is applied to a new climate region and crops, the model parameters need to be calibrated. Therefore, each crop should have its own N_c dilution curve model, considering the characteristics of the crop and the prevailing meteorological conditions [32,33].

Many studies have established the N_c dilution curve models based on aboveground biomass for different crops, including maize [34,35], rice [36], potato [37], and tomato [38]. The dry matter of sugar beet is readily available, and an N_c dilution curve model [28] can be used to calculate the current N concentration of the plant, which can be used to recommend the optimal N fertilization requirements. To address this, a three-year field experiment was conducted on sugar beet to establish an N_c dilution curve model of sugar beet under drip irrigation with plastic mulch for guiding sugar beet production in arid and semi-arid areas. Therefore, the objective of this study was to investigate the effects of irrigation level and N rate on sugar yield, N use efficiency, and soil nutrient residue, and to develop an Nc dilution curve model for sugar beet under drip irrigation in arid and semi-arid regions to support site-specific N management.

2. Materials and Methods

2.1. Experimental Site Description

A three-year field experiment was carried out on sugar beet in 2019, 2020, and 2021 at the test station of Thirty-One Group in Korla (40°53′ N, 86°56′ E; 900 m elevation), Xinjiang Province, Northwest China [39]. The experimental site is located in the continental desert climate region with an annual average precipitation of 59 mm and an annual mean temperature of 11 °C. The soil texture is Alfisol, which is sandy loam-textured with 2% clay, 38% silt, and 60% sand in the 0–80 cm soil layer, and the pH is 8.7. The bulk density is 1.61 g cm^{-3} and the water content at field capacity is 0.18 cm^3 cm^{-3}. The soil is slightly salinized with an average soil salt content of 1.07 g kg^{-1}. The experimental site had soil NO_3^--N of 13.79 mg kg^{-1}, available P of 6.52 mg kg^{-1}, and available K of 25.32 mg kg^{-1}. Details about meteorological data at the study site during the three years can be found in Figure S1.

2.2. Experimental Design

The N treatments consisted of 25, 60, 120, 240, 360, and 480 kg N ha^{-1} in the three years (designed as N_{25}, N_{60}, N_{120}, N_{240}, N_{360}, and N_{480}, respectively). The phosphorus and potassium rates were 120 kg P_2O_5 ha^{-1} and 60 kg K_2O ha^{-1}, respectively. Urea (46% N), ammonium dihydrogen phosphate (11% N; 53% P_2O_5), and potassium oxide (60% K_2O) were used as fertilizers. The irrigation amounts were based on fractions of crop evapotranspiration (ET_c) and combined with three irrigation coefficients (0.60, 0.80, and 1.00) in 2019 and 2020, which were denoted as $W_{0.6}$ (0.60 ET_c), $W_{0.8}$ (0.80 ET_c), and $W_{1.0}$ (1.00 ET_c), respectively. However, it was found that there was no significant difference in taproot yield between $W_{0.6}$ and $W_{0.8}$ in 2019 and 2020, and root yield showed a downward trend under $W_{1.0}$. Therefore, the three irrigation coefficients were adjusted to 0.40, 0.60, and 0.80, so the three irrigation amounts became $W_{0.4}$ (0.40 ET_c), $W_{0.6}$ (0.60 ET_c), and $W_{0.8}$ (0.80 ET_c) in 2021, respectively. ET_c was calculated as follows [40]:

$$ET_c = K_c \times ET_0 \tag{1}$$

where ET_0 is the reference crop evapotranspiration calculated according to the FAO-56 Penman–Monteith equation during the sugar beet growing seasons of 2019, 2020, and 2021 (Figure S1); K_c is the crop coefficient for sugar beet, which was determined according to the sugar beet growth stage. The K_c values used were 0.32 at the early stage, 0.97 at the middle stage, and 0.77 at the late stage [41]. ET_0 was calculated as follows [40]:

$$ET_0 = \frac{0.408\Delta(R_n - G) + \gamma\frac{900}{T_{mean}+273}u_2(e_s - e_a)}{\Delta + \gamma(1 + 0.34u_2)} \tag{2}$$

where ET_0 is the reference evapotranspiration (mm·day^{-1}), Δ is the slope of the saturation vapor pressure curve at air temperature (kPa·C^{-1}), R_n is the net radiation at the crop surface (MJ·m^{-2}·d^{-1}), G is the soil heat flux density (MJ·m^{-2}·d^{-1}), γ is the psychometric constant = ($0.665 \times 10^{-3} \times$ P), kPa·C^{-1} [40], u_2 is the wind speed at 2 m height (m·s^{-1}), e_s is the saturation vapor pressure (kPa), e_a is the actual vapor pressure (kPa); $(e_s - e_a)$ is the saturation vapor pressure deficit (kPa), and T_{mean} is the mean daily air temperature at 2 m height (°C).

Each treatment was set up in triplicate field plots and a randomized complete factorial block design (4.6 m wide and 12 m long). The sugar beet variety "*BETA356*" was planted on 12 April 2019, 13 April 2020, and 12 April 2021, respectively. The drip irrigation and fertilization system under plastic mulch included a water pump, filter, plastic film, fertilization tanks, and water distribution pipelines. The drip tape was an in-line drip tape with drippers spaced 30 cm apart. The dripper flow rate was 2.4 L h^{-1} and the dripper operating pressure was 0.1 MPa. The experimental area adopted the drip irrigation mode of one film, two belt lines with three beet lines, that is, two drip irrigation belts were laid under one film, and three lines of beets were planted. The width of the film was 106 cm and the width of the two bare areas between the films was 46 cm. Two drip irrigation belts were laid under the film at a distance of 30 cm from the edge of the film, and three rows of sugar beet were planted at a distance of 20 cm from the edge of the film and between the two drip irrigation belts, respectively. The planting spacing of sugar beet was 30 cm (Figure S2).

There were six fertilization events during the whole growth period (Figure S3). Fertilizers were applied before the middle stage of the sugar beet. The irrigation dates of each treatment were the same, and the irrigation interval was 7 days (Figure S3). The irrigation began in early June and stopped in late August. During sugar beet growing seasons, the actual irrigation amounts were 136 mm under $W_{0.4}$ in 2021; 186 mm, 198 mm, and 197 mm under $W_{0.6}$ in 2019, 2020, and 2021; 244 mm, 260 mm, and 258 mm under $W_{0.8}$ in 2019, 2020, and 2021; and 301 mm and 322 mm under $W_{1.0}$ in 2019 and 2020, respectively (Figure S3).

2.3. *Measurements and Methods*

2.3.1. Sample Analyses

Sugar Yield

All sugar beets of each plot were harvested, the leaves were removed, and the fresh taproots were washed, cleaned, and weighed to determine sugar beet taproot yield per plant and taproot yield per unit area. After calculating the taproot yield, three representative fresh taproots from each plot were further used to measure sucrose content, with a total of nine fresh taproots for each treatment. Sucrose content was measured with a saccharometer measure (PAL-1, Tokyo, Japan) (%) and the sugar yield per plant was determined. The sugar yield was further calculated as follows [42]:

$$\text{Sugar yield (t ha}^{-1}\text{)} = \text{taproot yield (t ha}^{-1}\text{)} \times \text{sucrose content (\%)} \quad (3)$$

According to Ebmeyer and Hoffmann [43], relative sugar yield is the sugar yield in relation to the grand mean of all N treatments in 2019, 2020 and 2021.

N Concentration and Uptake

Three representative plants were sampled to determine the total dry matter on 67, 89, 111, 137, and 161 days after planting in 2019, 78, 98, 118, and 160 days after planting in 2020, and 81, 101, 120, 140, and 162 days after planting in 2021 each plot, respectively. Fresh samples were immediately dried in an oven at 105 °C for 30 min for initial drying and subsequently dried at 75 °C to constant weight. After calculating the total dry matter, the samples were ground, extracted with H_2SO_4-H_2O_2, and analyzed for N concentration (g kg^{-1}) with a continuous-flow auto-analyzer. The total plant N uptake was further calculated as follows:

$$\text{Sugar yield (t ha}^{-1}\text{)} = \text{taproot yield (t ha}^{-1}\text{)} \times \text{sucrose content (\%)} \quad (4)$$

N Use Efficiency

According to Ebmeyer and Hoffmann [43], N uptake efficiency (NU$_P$E), N utilization efficiency (NU$_t$E), and N use efficiency (NUE) of sugar beet were calculated as follows:

$$NU_pE = \frac{\text{N uptake (kg ha}^{-1}\text{)}}{\text{N rate (kg ha}^{-1}\text{)}} \quad (5)$$

$$NU_tE = \frac{\text{Sugar yield (kg ha}^{-1}\text{)}}{\text{N uptake (kg ha}^{-1}\text{)}} \quad (6)$$

$$NUE = \frac{\text{Sugar yield (kg ha}^{-1})}{\text{N rate (kg ha}^{-1})} \quad (7)$$

Soil NO_3-N Residue

Ammonium nitrogen (NH_4-N) is easily absorbed by crops and not easily leached in the deep soil [44], so we have mainly focused on the content of nitrate–nitrogen (NO_3-N) in the soil. After the sugar beet was harvested, soil samples were taken by the soil drilling method. Samples at four horizontal locations (0, 20, 30, and 53cm from the edge of the plastic film in the 0–20 cm, 20–40cm, 40–60 cm, and 60–80 cm soil layers were collected) (Figure S2). The soil samples were subjected to air dry and fine screening, 2 mol L^{-1} Potassium Chloride (*KCl*) solution was then used for extraction (dry soil 5 g, soil liquid ratio 1:10), and the content of NO_3-N in the soil was determined by flow analyzer.

2.3.2. Critical Nitrogen (N_c) Dilution Curve

N_c Dilution Curve Based on Dry Matter

The N_c dilution curve model was used to evaluate the N nutrition status of sugar beet. The N_c dilution curve was calculated as follows [28]:

$$N_c = a\, DM_{max}^{-b} \quad (8)$$

where N_c is the N_c concentration (g kg^{-1}); a is the N concentration for DM = 1 Mg ha^{-1} [32]; DM_{max} is the highest dry matter at the minimum N rate (Figure S4, Mg ha^{-1}); b is dimensionless and represents the ratio between the relative decline in plant N concentration and the relative crop growth rate [45]. The dry matter and N concentration samples in 2020 and 2021 (n = 5, 9, 9, and 4 under $W_{0.4}$, $W_{0.6}$, $W_{0.8}$, and $W_{1.0}$, respectively) were used for model development since they included all irrigation treatments, and the dry matter and N concentration in 2019 (n = 5, 5 and 5 under $W_{0.6}$, $W_{0.8}$, and $W_{1.0}$, respectively) was used for model validation (Figure S4).

Evaluation of Model Performance

Root Mean Square Error (RMSE) and Normalized Root Mean Square Error (n-RMSE) were used to validate the model, and a 1:1 scatter plot between the simulated N_c value and the measured value was used to ascertain the reliability of the model. The RMSE and n-RMSE were calculated as [1,29]:

$$\text{RMSE} = \sqrt{\frac{\sum_{i=1}^{n}(S_i - M_i)^2}{n}} \quad (9)$$

$$\text{n} - \text{RMSE} = \frac{\text{RMSE}}{\overline{m}} \times 100\% \quad (10)$$

where S_i is the simulated value; M_i is the measured value; n is the data number; $\overline{m}$ is the average of observed values. Generally, with n-RMSE < 10%, a model is considered to be extremely stable; 10% < n-RMSE < 20% is considered stable; 20% < n-RMSE < 30% is considered generally stable; n-RMSE > 30% is considered to have poor stability.

Nitrogen Nutrition Index

The N nutrition index (NNI) is an important indicator to determine the N nutritional status of crops [46,47]. The NNI was calculated as follows [48]:

$$\text{NNI} = \frac{N_i}{N_c} \quad (11)$$

where N_i is the actual N concentration (%). Values of NNI < 1 indicate that biomass growth was limited by the N supply, while a value of NNI = 1 indicates that biomass growth was not limited by the N supply, and values of NNI > 1 indicate crops under luxury N conditions.

2.4. Data Analysis and Statistics

Data were compiled and analyzed using Excel 2010 (Microsoft Corporation, Albuquerque, NM, USA). SPSS 24.0 was used for statistical analyses (SPSS Inc. Chicago, IL, USA). Origin 2018 was used to create the figures. Before statistical analysis, we first assessed the normality (Shapiro–Wilk test) and constant variance (computing the Spearman rank correlation between the absolute values of the residuals and the observed values of HA) of the data. The two-way (irrigation by N) analysis of variance (ANOVA) was used to assess the treatment effects on sugar yield, N uptake, NU_PE, NU_tE and NUE in each experimental year. In order to calculate the N_c concentration values, the dry matter values in different N rates under each irrigation amount in each experimental year were compared using one-way ANOVA (SPSS Inc. Chicago, IL, USA) by Tukey's HSD test at the 5% level ($p < 0.05$). All pairwise comparisons of the treatment means were performed using ANOVA (SPSS Inc. Chicago, IL, USA) by Tukey's HSD test at the 5% level ($p < 0.05$).

3. Results

3.1. Sugar Yield, N Uptake, Use Efficiency, and Residual

3.1.1. Sugar Yield

The irrigation–nitrogen (I × N) interaction had no significant effect on sugar yield in 2019 ($p > 0.05$), but it had significant effects on sugar yield in 2020 and 2021 ($p < 0.05$). For the three years, the sugar yield varied from 7.6 to 12.5 Mg ha^{-1} in 2019, from 7.8 to 12.5 Mg ha^{-1} in 2020, and from 5.8 to 12.3 Mg ha^{-1} in 2021, respectively. The sugar yield increased and then generally stabilized with the increasing N rate under the same irrigation amount in the same year (Figure 1a–c).

3.1.2. N Uptake

The I × N interaction had a significant effect on N uptake in 2019 ($p < 0.05$), but it had no significant effect on N uptake in 2020 and 2021 ($p > 0.05$). For the three years, the total N uptake varied from 145.5 to 364.2 kg ha^{-1} in 2019, from 161.8 to 398.8 kg ha^{-1} in 2020, and from 86.2 to 347.8 Mg ha^{-1} in 2021, respectively. In general, the N uptake increased with the increasing N rate under the same irrigation amount. However, most treatments showed little difference in N uptake under the same N rate in the same year (Figure 1d–f).

3.1.3. NU_PE, NU_tE and NUE

The I × N interaction had no significant effect on NU_PE in the three years ($p > 0.05$), but it had a significant effect on NU_tE and NUE in the three years ($p < 0.05$). For the three years, the NU_PE varied from 0.7 to 6.1 kg kg^{-1} in 2019, from 0.8 to 7.2 kg kg^{-1} in 2020, and from 0.7 to 4.0 kg kg^{-1} in 2021, respectively. The NU_tE varied from 26.0 to 60.6 kg kg^{-1} in 2019, from 29.2 to 57.3 kg kg^{-1} in 2020, and from 28.0 to 101.7 kg kg^{-1} in 2021, respectively. The NUE varied from 18.9 to 359.0 kg kg^{-1} in 2019, from 22.4 to 369.1 kg kg^{-1} in 2020, and from 20.3 to 369.7 kg kg^{-1} in 2021, respectively. The trends of NU_PE, NU_tE, and NUE are roughly the same among the treatments, that is, the maxima of NU_PE, NU_tE, and NUE were observed at N_{25}, and the minima of NU_PE, NU_tE, and NUE were observed at N_{480} for a given irrigation amount (Figure 2).

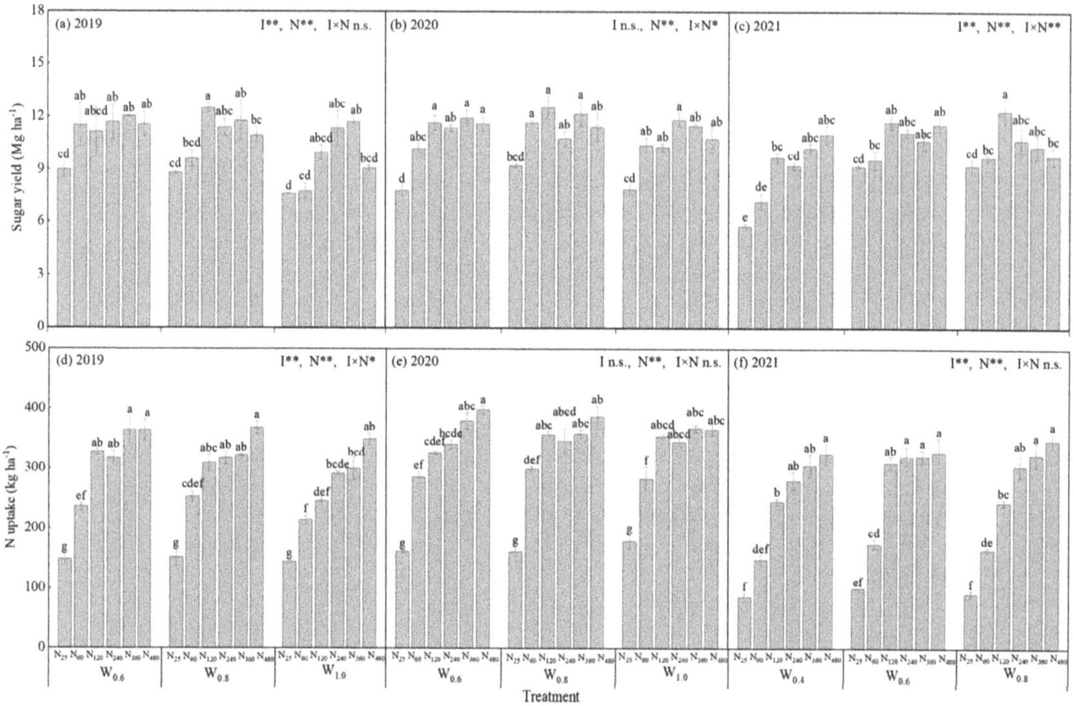

Figure 1. Sugar yield and nitrogen (N) uptake in different irrigation and nitrogen treatments in 2019, 2020, and 2021. (**a**) Sugar yield in 2019; (**b**) Sugar yield in 2020; (**c**) Sugar yield in 2021; (**d**) N uptake in 2019; (**e**) N uptake in 2020; and (**f**) N uptake in 2021. Bars are the means ± one standard error of the mean (n = 3). * and ** mean a significant difference at $p < 0.05$ and $p < 0.01$, respectively; n.s. means no significant difference ($p > 0.05$). Different letters indicate the significant difference between treatments at 0.05 level by Tukey's HSD test in the same year. $W_{0.4}$, $W_{0.6}$, $W_{0.8}$, and $W_{1.0}$ are irrigation amounts of 0.4, 0.6, 0.8, and 1.0 ET_c, respectively. N_{25}, N_{60}, N_{120}, N_{240}, N_{360}, and N_{480} are nitrogen rates of 25, 60, 120, 240, 360, and 480 kg N ha^{-1}, respectively.

3.1.4. Soil NO_3-N Distribution in the 0–80 cm Soil Profile at Harvest

There was no significant difference in soil NO_3-N contents among N_{25}, N_{60}, and N_{120} under the same irrigation amount, and their soil NO_3-N contents were lower. In general, soil NO_3-N contents increased as N rates increased, and this phenomenon was more obvious with the higher N rates. By calculating the average soil NO_3-N residual amount in the 0–80 cm soil layer of each irrigation amount during the three years, we found that the maximum average soil NO_3-N accumulation was obtained at the $W_{0.4}$, and the minimum was obtained at the $W_{1.0}$. In the horizontal distance, with the drip line as the centre, soil NO_3-N moved to both sides with water. Many treatments had a soil NO_3-N accumulation zone, which was located on the soil surface layer or at a depth of 40 cm below the soil surface. This phenomenon became more obvious with the increase in the N rate. In addition, the soil NO_3-N of $W_{0.8}N_{480}$ and $W_{1.0}N_{480}$ in 2019 and $W_{1.0}N_{480}$ in 2020 moved into deep soil layers, but that of other treatments was mainly distributed in the 0–60 cm soil layer in the three years (Figure 3; Figure S7).

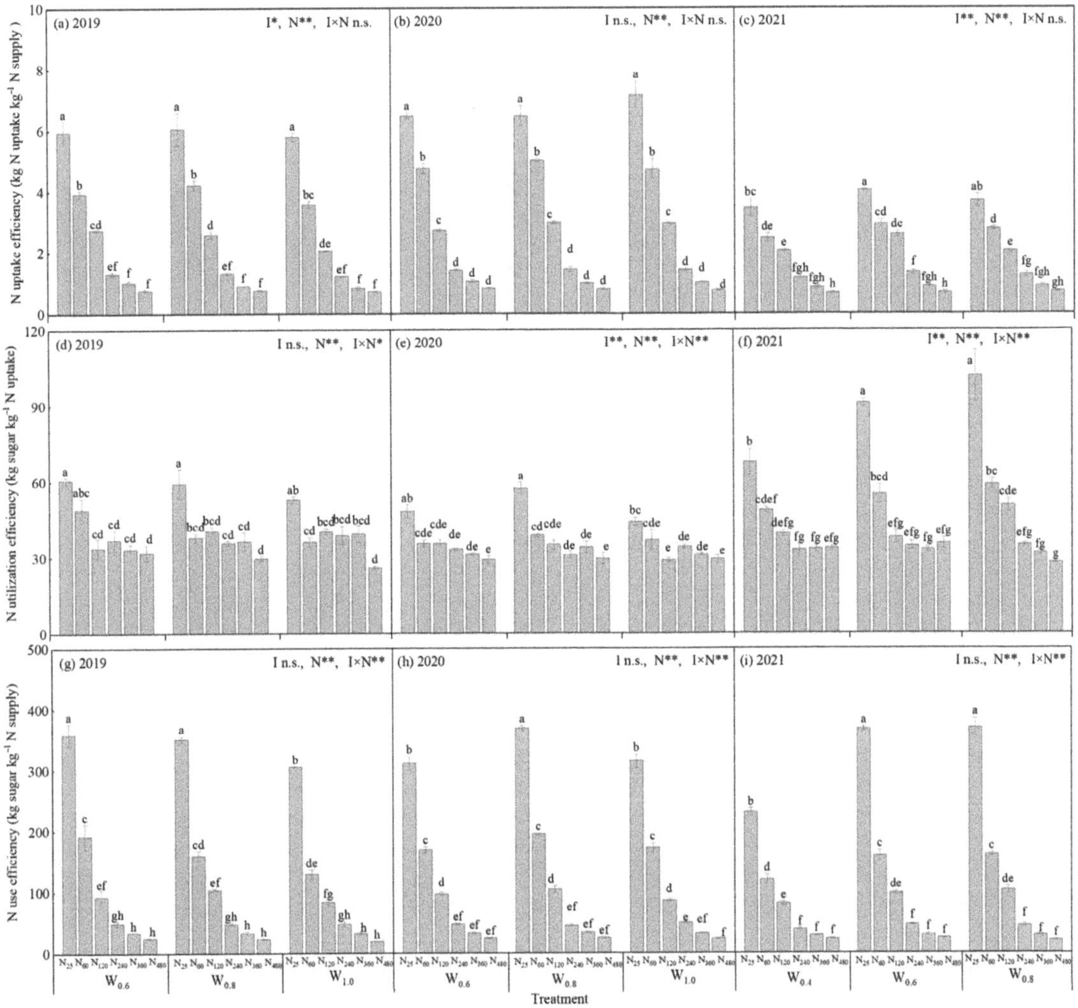

Figure 2. Nitrogen (N) uptake efficiency, utilization efficiency, and use efficiency of sugar beet in different irrigation and nitrogen treatments in 2019, 2020, and 2021. Bars are the means ± one standard error of the mean (n = 3). **(a)** N uptake efficiency in 2019; **(b)** N uptake efficiency in 2020; **(c)** N uptake efficiency in 2021; **(d)** N utilization efficiency in 2019; **(e)** N utilization efficiency in 2020; **(f)** N utilization efficiency in 2021; **(g)** N use efficiency in 2019; **(h)** N use efficiency in 2020; and **(i)** N use efficiency in 2021. * and ** mean a significant difference at $p < 0.05$ and $p < 0.01$, respectively; n.s. means no significant difference ($p > 0.05$). Different letters indicate the significant differences in the treatments at 0.05 level by Tukey's HSD test in the same year. $W_{0.4}$, $W_{0.6}$, $W_{0.8}$, and $W_{1.0}$ are irrigation amounts of 0.4, 0.6, 0.8, and 1.0 ET_c, respectively. N_{25}, N_{60}, N_{120}, N_{240}, N_{360}, and N_{480} are nitrogen rates of 25, 60, 120, 240, 360, and 480 kg N ha^{-1}, respectively.

3.2. *N_c Dilution Curve Model*

3.2.1. Determination of Nc Dilution Curve Model

As shown in Figures S4 and S5, we found that increasing the N rate increased the dry matter and N concentration of sugar beet. In addition, the dry matter of sugar beet increased with the increase in days after planting, and the N concentration decreased as the days after planting increased.

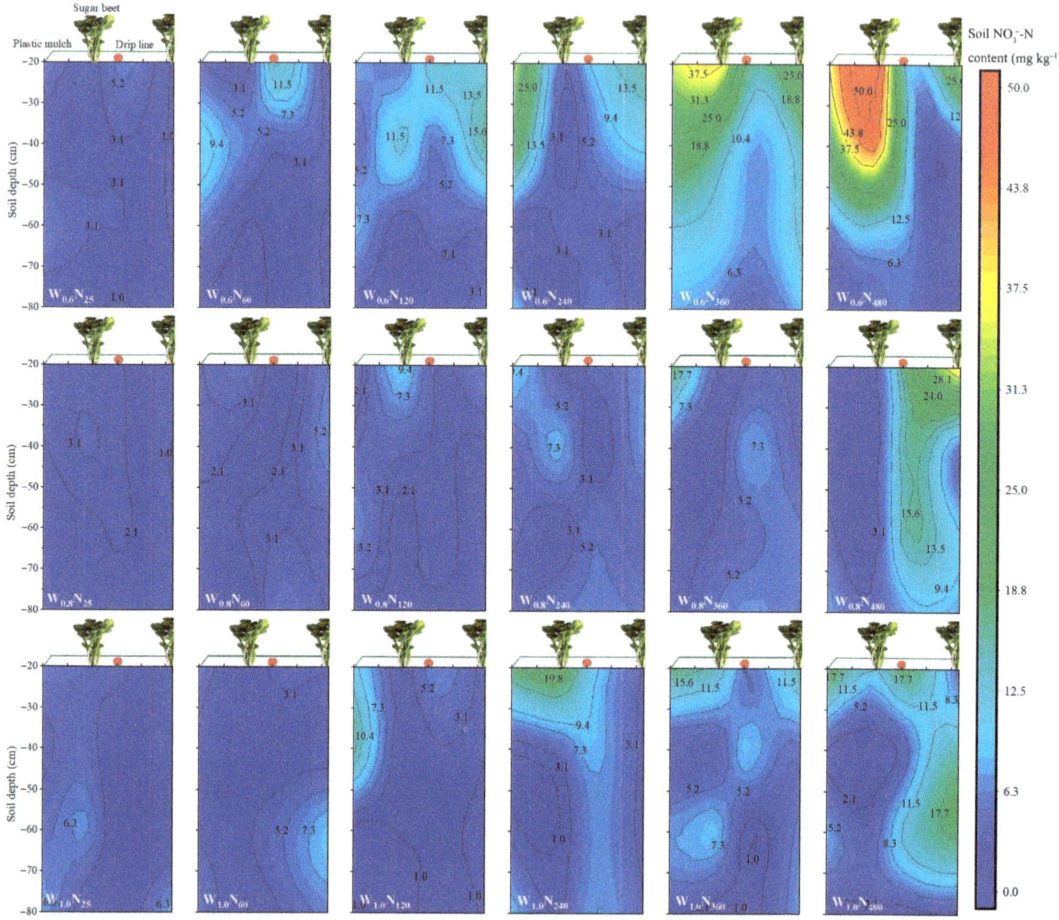

Figure 3. Examples of the spatial distribution of soil NO_3^--N in each treatment after the experiment in 2019. $W_{0.6}$, $W_{0.8}$, and $W_{1.0}$ are irrigation amounts of 0.6, 0.8, and 1.0 ET_c, respectively. N_{25}, N_{60}, N_{120}, N_{240}, N_{360}, and N_{480} are nitrogen rates of 25, 60, 120, 240, 360, and 480 kg N ha^{-1}, respectively. The spatial distribution of soil NO_3^--N in each treatment after the experiment in 2020 and 2021 can be found in Figure S7.

Before selecting data points, one-way ANOVA was carried out on the dry matter of sugar beet of different N rates under each irrigation amount in the three years. Data points from each sampling from 78 to 160 days after planting in 2020 and from 81 to 162 days after planting in 2021 were used to determine the N_c concentration points by following the method proposed by [28]. The N_c dilution curve models with different irrigation amounts all showed good performance in the calibration dataset ($p < 0.001$, Figure 4). In addition, we found that model parameters a and b had significant quadratic correlations with the irrigation coefficients. Therefore, a and b were fitted linearly with the corresponding irrigation coefficients. The fitting equations were as follows:

$$a = 25.06 \times I^2 - 42.64 \times I + 48.28, R^2 = 0.90, p < 0.001 \tag{12}$$

$$b = -0.07 \times I^2 - 0.07 \times I + 0.38, R^2 = 0.99, p < 0.001 \tag{13}$$

where I is the irrigation coefficient. Therefore, the N_c dilution curves of sugar beet under different irrigation amounts can be expressed as follows:

$$N_c = \left(25.06 \times I^2 - 42.64 \times I + 48.28\right) \times DM^{-(-0.07 \times I^2 - 0.07 \times I + 0.38)}, R^2 = 0.91, p < 0.001 \tag{14}$$

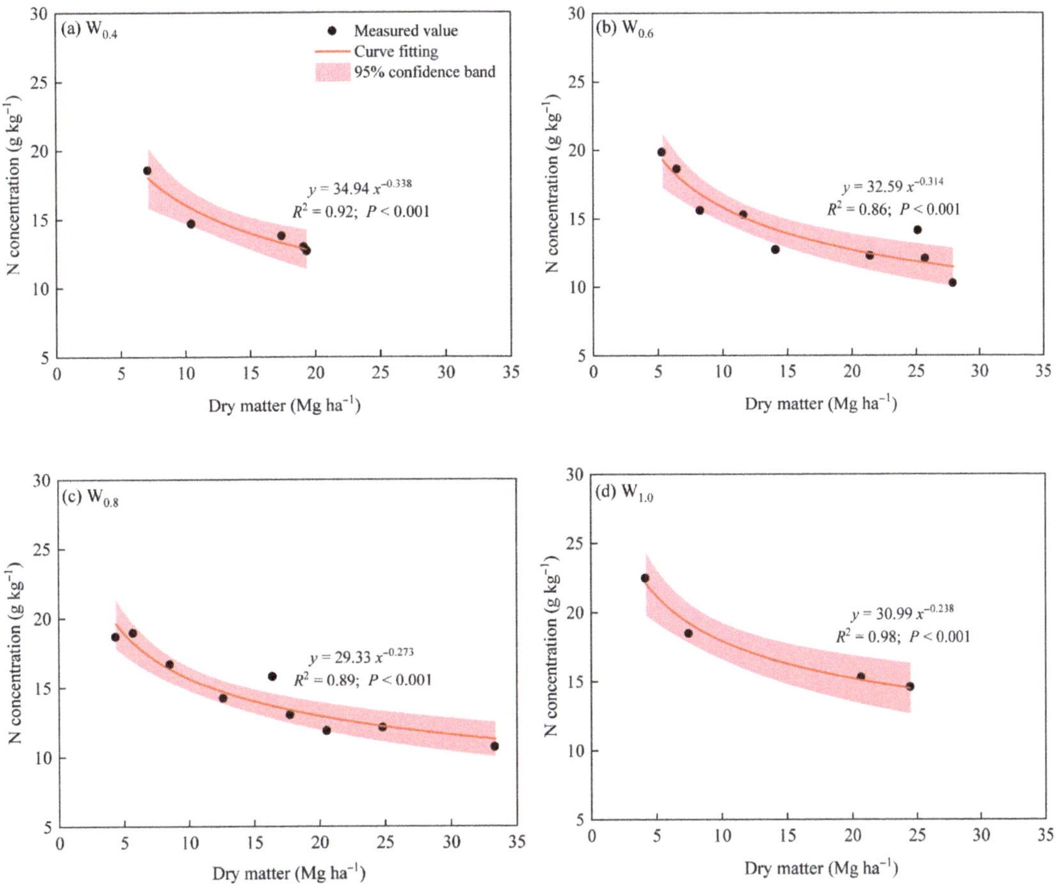

Figure 4. Critical nitrogen dilution curves for sugar beet under different irrigation amounts. (**a**) critical nitrogen dilution curves under $W_{0.4}$ irrigation amount; (**b**) critical nitrogen dilution curves under $W_{0.6}$ irrigation amount; (**c**) critical nitrogen dilution curves under $W_{0.8}$ irrigation amount; and (**d**) critical nitrogen dilution curves under $W_{1.0}$ irrigation amount. $W_{0.4}$, $W_{0.6}$, $W_{0.8}$, and $W_{1.0}$ mean irrigation amounts of 0.4, 0.6, 0.8, and 1.0 ET_c, respectively.

3.2.2. Validation of N_c Dilution Curve Model

The validation of the model was carried out using the data from 2019 under different irrigation levels ($W_{0.6}$, $W_{0.8}$, and $W_{1.0}$). As shown in Figure 5, the model accurately estimated the N_c concentration with an RMSE of 2.50 g kg^{-1} and n-RMSE of 16.6%. The n-RMSE value was more than 10% but less than 20%, indicating that the established model was stable and could be further used for the diagnosis of N nutrition in the sugar beet growth period.

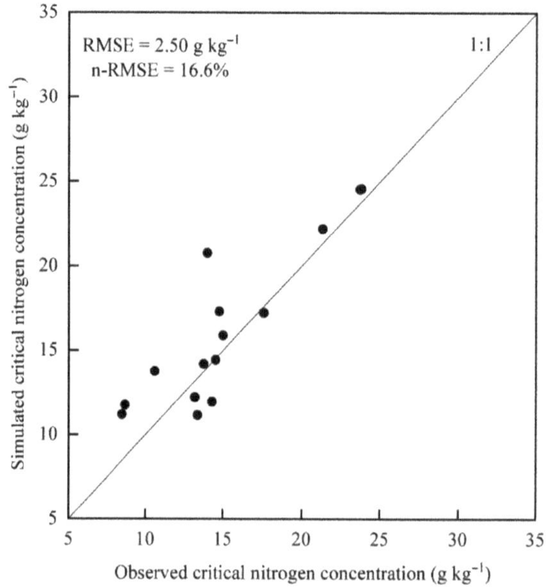

Figure 5. Comparison of simulated and observed critical nitrogen concentrations. Each black dot represents a paired data point of observed versus simulated critical nitrogen concentration.

3.2.3. N Nutrition Diagnosis Based on the Nc Dilution Curve Model

As shown in Figure S6, sugar beet is more prone to N deficit at the early growth stage (*NNI* < 1). In general, *NNI* showed an increasing trend with the increase in the N rate. For a given irrigation amount, *NNI* under N_{25} and N_{60} were all less than 1. At harvest, the *NNI* of most treatments were less than 1.

3.2.4. Relationship Between Relative Sugar Yield and NNI

The relationships between relative sugar yield (*RY*) and *NNI* were expressed by linear-plateau function in different irrigation amounts (Figure 6). The regression models were:

$$W_{0.4}: RY = \begin{cases} -1.06 + 2.09\, NNI, & NNI < 0.98 \\ 0.98, & NNI \geq 0.98 \end{cases} \quad R^2 = 0.82,\ p < 0.01 \tag{15}$$

$$W_{0.6}: RY = \begin{cases} -1.33 + 2.51\, NNI, & NNI < 0.97 \\ 1.10, & NNI \geq 0.97 \end{cases} \quad R^2 = 0.99,\ p < 0.01 \tag{16}$$

$$W_{0.8}: RY = \begin{cases} -0.19 + 1.34\, NNI, & NNI < 0.90 \\ 1.02, & NNI \geq 0.90 \end{cases} \quad R^2 = 0.70,\ p = 0.01 \tag{17}$$

$$W_{1.0}: RY = \begin{cases} -0.61 + 1.75\,NNI, & NNI < 0.88 \\ 0.93, & NNI \geq 0.88 \end{cases} \quad R^2 = 0.93,\ p < 0.01 \tag{18}$$

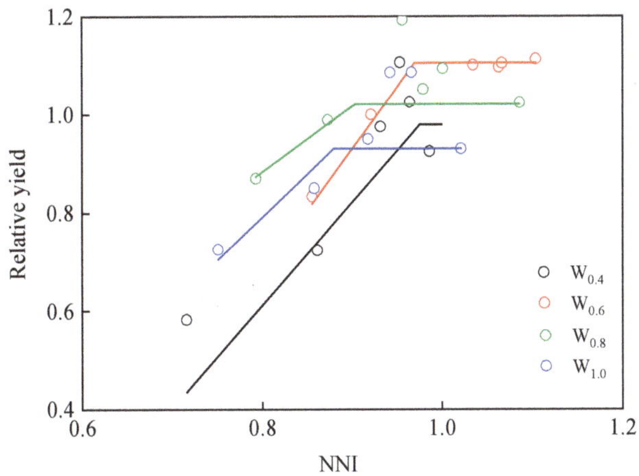

Figure 6. Relationship between relative sugar yield and nitrogen nutrition index (*NNI*) for four irrigation amounts and six nitrogen rates. The *NNI* data were averaged for N_{25}, N_{60}, N_{120}, N_{240}, N_{360}, and N_{480} treatments over three years. Colored lines represent different irrigation levels: black for $W_{0.4}$, red for $W_{0.6}$, green for $W_{0.8}$, and blue for $W_{1.0}$.

Based on the relationships, for NNI = 0.97 of $W_{0.6}$, the RY was highest, while for NNI > 0.97 of $W_{0.6}$, the RY was generally stable at 1.10.

4. Discussion

4.1. Effects of Irrigation Amount and N Rate on Sugar Yield, N Uptake and Use Efficiency, and Soil NO_3-N Residue

4.1.1. Sugar Yield

The significant irrigation–nitrogen interaction on sugar yield of sugar beet observed in 2020 and 2021 was primarily driven by irrigation amount under N_{25} and N_{60}, as well as variable responses to increasing N from 25 to 120 kg ha^{-1}. For a given irrigation amount, we found that sugar yield increased and then roughly stabilized as the N rate increased. This aligns with findings from Kiymaz and Ertek [12], who reported optimal sugar yield at specific irrigation and N rates. The decline or stabilization in yield at higher N rates may be attributed to reduced sucrose content and increased impurities (Na, K, and amino-N) [12]. Similarly, reduced irrigation led to lower sugar yield due to more water accumulation in the taproot, which decreases sucrose content [21,49,50]. Therefore, controlled irrigation amount and N rate for increasing sugar yield are of vital importance.

4.1.2. N Uptake and Use Efficiency

Analyzing crops' N uptake and use efficiency can help to evaluate whether the N rate is suitable [9]. In this study, total N uptake was higher than observed in other studies [9,42,43], likely due to a combination of increased N supply, crop nutrient demand, and irrigation method. Some studies reported that plant N uptake was determined by both the plant N rate and N demand [43,51]. Thus, the treatment with higher N uptake can be explained by a higher growth rate and N rate (Figure S4). In addition, except for the different experimental weather conditions and N rates among the studies, another possible reason is that we used drip irrigation technology in our study. This technology has been demonstrated to improve plant nutrient uptake and yield [52], and it is widely used for saving water and fertilizer resources in agricultural production in arid and semi-arid regions.

We found that plant total N uptake showed opposite trends to NU_pE, NU_tE, and NUE with the increasing N rate. This inverse relationship is consistent with Armstrong et al. [51], who observed that sugar beet can exhaust the N from the soil during the growth period. Thus, the sugar beet under a lower N supply further exhausted the N in the soil and increased NU_pE. The lower soil NO_3-N residue content under the lower N supply treatment supports this explanation. The NU_pE, NU_tE, and NUE had a similar trend as observed for sugar beet under varying N rates under the same irrigation amounts. The high NU_tE indicated effective remobilization of N under the lower N supply [53]. However, a lower N supply did not increase the sugar yield in the three years. In terms of pursuing a high sugar yield, this N rate (e.g., N_{25}) is not desirable.

4.1.3. Soil NO_3-N Residue

In this study, soil NO_3-N content and distribution were significantly affected by irrigation amount, N rate, and crop N uptake was also reported by Gorska et al. [54]. Excessive N input not only decreased N uptake and utilization efficiency but also increased the soil NO_3-N residue, thereby increasing the risk of environmental pollution [55,56]. The irrigation–nitrogen interaction notably influenced the distribution of soil NO_3-N. Increasing irrigation amount may decrease the soil NO_3-N content in the 0–80 cm soil layer, likely due to enhanced nitrate leaching or deeper movement. These results concur with Badr et al. [57], who found that the soil NO_3-N was readily soluble in water and moved with the water. Additionally, the root system of crops causes more water to move in the horizontal direction after entering the soil and accumulates on the edge of the wetting front. From a mechanistic perspective, drip fertigation improves NUE by synchronizing N supply with crop demand and reducing nitrate buildup beyond the root zone. Therefore, we should choose appropriate water and N management to meet the growing needs of crops, reduce waste of water and fertilizer resources, and ultimately achieve high yields.

4.2. *N_c Dilution Curve Model*

4.2.1. Nc Dilution Curve Model Based on Dry Matter

In the current study, the selected data points of dry matter used to establish the N_c dilution curve were > 1 Mg ha^{-1} [28]. Despite this, the selected data points had a shortcoming in that most of the data points were collected when the dry matter yield was more > 2 Mg ha^{-1}. Further studies should consider some data points of dry matter above but close to 1 Mg ha^{-1}, to better verify the acceptability of the N_c dilution curve.

In agricultural practice, the N_c dilution curve is an important diagnostic tool to assess the N profit and loss status of crops at the growth stage; as long as the dry matter of crops is available, the N nutrition status of crops can be determined according to the N_c dilution curve [58,59]. For example, if the measured N concentration is on the N_c dilution curve, it indicates that the N application rate was optimal. On the contrary, if the measured N concentration is higher (or lower) than the N_c dilution curve, it indicates that there is an N surplus (or deficiency) in the plant [29,60]. However, there were some differences in the model parameters between our study and the previous studies. Parameter comparisons revealed noticeable differences between our model and previous studies. Firstly, sugar beet is a C3 crop, but we found that parameter a in the present study was lower than that of Greenwood et al. [31], who established a general N_c dilution curve for C3 crops (a = 57.0 g kg^{-1}). Indeed, previous studies have shown that the parameters of the N_c dilution curve are affected by many factors, including agronomic practices, the ecological climate of crop growth, crop types, and varieties [37,38,58]. For example, the parameters a (43.6 g kg^{-1}) of the N_c dilution curve of winter wheat in the Guanzhong Plain in China [29] were lower than the parameters a (53.5 g kg^{-1}) of the N_c dilution curve of winter wheat in France [28]. Potato and tomato belong to the C3 group of crops, but the parameters a are not the same between potato (a = 53.0 g kg^{-1}) and tomato (a = 45.3 g kg^{-1}) [37,38]. It is indicated that N_c dilution curve parameters vary with crop species, cultivars, management practices, and climate conditions. Therefore, it is expected that each crop needs its model parameters [32]. Secondly, the value of parameters b in this study was different from that of other studies, and one of the main reasons was that the crop types were different [28,37,38,58]. For example, Qiang et al. [29] reported that the parameter b was higher in C3 species than in C4 species when the N_c dilution curve was based on dry matter. In addition, the parameters b indicate the decline in plant N concentration with crop growth and depend on plant N uptake relative to dry matter increase. This reflects physiological differences in nitrogen uptake and biomass allocation between crop types. Therefore, the decrease in plant N concentration during vegetative growth can be ascribed to the decreased N concentration per unit of dry matter.

4.2.2. Optimal Irrigation Amount and N Rate Based on the Diagnosis of N Nutrition Status

In this study, the main purpose of establishing the N_c dilution curve was to diagnose the N nutritional status of crops by agronomic research methods, which have been applied widely in agricultural production [34,58]. Thus, real-time diagnosis of plant N nutrition status based on NNI is the basis for rational N application, promoting crop growth and increasing economic yield, and NNI is a reliable diagnostic tool for evaluating the N status of crops [26,48].

This study analyzed the variation range of NNI during the growth period of sugar beet under different water and N supplies and obtained the appropriate N application rate under different irrigation conditions for three years. We found that sugar beet was more prone to N deficit at the early stage of growth. Meanwhile, it was reported that maintaining soil N deficiency for 42 days before harvest improves sugar beet quality [61]. To improve the N use efficiency, absorption, and utilization, N fertilizer should be applied at the early growth stage of sugar beet, and there should be no excess N in the soil 4–6 weeks before harvest [22].

We found that the relationships between relative sugar yield (RY) and NNI were expressed by linear-plateau functions and accounted for more than 70% of the variation. A similar relationship was also found between RY and NNI in potatoes [37], maize [62], and greenhouse cherry tomato [63]. The relationships confirmed the link between N uptake and sugar yield of sugar beet, confirming the importance of NNI for predicting actual sugar yield with respect to potential sugar yield. In addition, for an NNI > 0.97 of $W_{0.6}$, the RY was generally stable at 1.10 and higher than that of other irrigation amounts, which indicated that the optimal NNI was 0.97 under $W_{0.6}$. Meanwhile, since the agricultural water resources of each sugar beet planting area are different, it is important to choose the appropriate N rate that corresponds to the actual irrigation amount and reduce the waste of water and N fertilizer resources.

5. Conclusions

This study demonstrated that optimizing irrigation and nitrogen (N) input significantly improved sugar beet yield and nitrogen use efficiency (NUE) under arid and semi-arid conditions. The maximum relative sugar yield (RY = 1.10) was achieved at an

N nutrition index of 0.97 under 0.6 ETc irrigation, indicating that moderate irrigation coupled with proper N supply is optimal for both productivity and environmental protection. Although plant N uptake and concentration increased with N rate, higher NUE, NU_PE, and NU_tE were observed under lower N inputs. Excessive N application (480 kg ha^{-1}) resulted in NO_3^-–N accumulation below 60 cm soil depth, implying increased leaching risk. A crop-specific Nc dilution curve based on dry matter was developed with a stable fit (n-RMSE = 16.6%), and its parameters showed significant quadratic responses to irrigation level. Compared to curves established for other C_3 crops and regions, the differences in parameter values highlighted the need for regionally adapted Nc models. Overall, the Nc curve approach proves effective for diagnosing sugar beet N status and provides a valuable tool for improving N management in similar agro-ecological zones.

Supplementary Materials: The following supporting information can be downloaded at: https://www.mdpi.com/article/10.3390/plants14132055/s1, Figure S1: (a) Daily average air temperature, (b) precipitation, (c) solar radiation and (d) reference crops evapotranspiration (ET_0) during growing seasons of sugar beet in 2019, 2020, and 2021; Figure S2: Layout of sugar beet drip irrigation under plastic mulch and soil sampling locations; Figure S3: Timing and amounts of drip irrigation under plastic mulching and the proportions of total fertilizer applied each time for sugar beet in (a) 2019, (b) 2020 and (c) 2021. Figure S4: The changes of aboveground biomass over time under different treatments in 2019, 2020 and 2021. Different letters indicate the significance on the same day after planting between different treatments at 0.05 level by Tukey's HSD test. $W_{0.4}$, $W_{0.6}$, $W_{0.8}$ and $W_{1.0}$ are irrigation amounts of 0.4, 0.6, 0.8 and 1.0 ET_c, respectively. N_{25}, N_{60}, N_{120}, N_{240}, N_{360} and N_{480} are nitrogen rates of 25, 60, 120, 240, 360 and 480 kg N ha^{-1}, respectively. The black labels in 2020 and 2021are the selected calibration dataset, and the black labels in 2019 are the selected validation dataset. Figure S5: The changes in nitrogen concentration of dry matter over time under different treatments in 2019, 2020 and 2021. Bars are the means ± one standard error of the mean (n = 3). $W_{0.4}$, $W_{0.6}$, $W_{0.8}$ and $W_{1.0}$ are irrigation amounts of 0.4, 0.6, 0.8 and 1.0 ET_c, respectively. N_{25}, N_{60}, N_{120}, N_{240}, N_{360} and N_{480} are nitrogen rates of 25, 60, 120, 240, 360 and 480 kg N ha^{-1}, respectively. Figure S6: The dynamics of nitrogen nutrition index (NNI) over time under different irrigation amounts. The blue dashed line represents the NNI = 1, which means that sugar beet is under optimal nitrogen nutrition status. $W_{0.4}$, $W_{0.6}$, $W_{0.8}$ and $W_{1.0}$ are irrigation amounts of 0.4, 0.6, 0.8 and 1.0 ET_c, respectively. N_{25}, N_{60}, N_{120}, N_{240}, N_{360} and N_{480} are nitrogen rates of 25, 60, 120, 240, 360 and 480 kg N ha^{-1}, respectively. Figure S7: Spatial distribution of Soil NO_3^--N in each treatment after the experiment in 2020 (a) and 2021 (b). $W_{0.4}$, $W_{0.6}$, $W_{0.8}$ and $W_{1.0}$ are irrigation amounts of 0.4, 0.6, 0.8 and 1.0 ET_c, respectively. N_{25}, N_{60}, N_{120}, N_{240}, N_{360} and N_{480} are nitrogen rates of 25, 60, 120, 240, 360 and 480 kg N ha^{-1}, respectively.

Author Contributions: Conceptualization, Y.W., F.Y. and F.Z.; Methodology, F.Y.; Software, Y.W. and F.Z.; Validation, F.Y. and J.F.; Formal analysis, Y.W.; Investigation, Y.W. and J.F.; Writing—original draft, Y.W.; Writing—review and editing, Y.W., F.Y., J.F. and F.Z.; Visualization, J.F.; Supervision, F.Y. and F.Z.; Project administration, F.Z.; Funding acquisition, F.Z. All authors have read and agreed to the published version of the manuscript.

Funding: This study was jointly supported by the National Key Research and Development Program of China (No. 2017YFC0403303), the National Natural Science Foundation of China (No. 51979231), and the "111" Project (No. B12007).

Data Availability Statement: Data are contained within the article and Supplementary Materials.

Conflicts of Interest: The authors declare no conflict of interest.

References

1. Hou, X.; Xiang, Y.; Fan, J.; Zhang, F.; Hu, W.; Yan, F.; Guo, J.; Xiao, C.; Li, Y.; Cheng, H.; et al. Evaluation of cotton N nutrition status based on critical N dilution curve, N uptake and residual under different drip fertigation regimes in Southern Xinjiang of China. *Agric. Water Manag.* **2021**, *256*, 107134. [CrossRef]
2. Zhou, H.; Wang, L.; Xu, P.; Liu, D.; Hao, Y.; Wang, K.; Fan, H. Increasing nitrogen supply during limited irrigation improves drip-irrigated sugar beet growth, yield, and net return in arid areas. *Field Crops Res.* **2024**, *299*, 109646. [CrossRef]
3. Nassah, H.; Zitouni, A.; Tibssirte, O.; Daghor, L.; Chatoui, H.; Midhat, L. Irrigation water needs analysis and management strategies in the Tadla irrigation zone: A case study of sugar beet plots. *Ecol. Eng. Environ. Technol.* **2025**, *26*, 200563. [CrossRef]
4. Cui, Z.; Chen, X.; Miao, Y.; Zhang, F.; Sun, Q. On-farm evaluation of the improved soil N_{min}-based nitrogen management for summer maize in North China Plain. *Agron. J.* **2008**, *100*, 517–525. [CrossRef]
5. Li, Y.; Liu, H.; Huang, G.; Zhang, R.; Yang, H. Nitrate nitrogen accumulation and leaching pattern at a winter wheat: Summer maize cropping field in the North China Plain. *Environ. Earth Sci.* **2016**, *75*, 118. [CrossRef]
6. Mainardis, M.; Cecconet, D.; Moretti, A.; Callegari, A.; Goi, D.; Freguia, S.; Capodaglio, A. Wastewater fertigation in agriculture: Issues and opportunities for improved water management and circular economy. *Environ. Pollut.* **2022**, *296*, 118755. [CrossRef]
7. Brye, K.R.; Norman, J.M.; Gower, S.T.; Bundy, L.G. Methodological limitations and N-budget differences among a restored tallgrass prairie and maize agroecosystems. *Agric. Ecosyst. Environ.* **2003**, *97*, 181–198. [CrossRef]
8. Su, J.; Zhou, H.; Wang, K.; Fan, H.; Hou, Z. Effects of nitrogen fertilizer management on dry matter accumulation and yield of drip-irrigated sugar beet in arid areas. *Agronomy* **2024**, *14*, 1010. [CrossRef]
9. Laufer, D.; Nielsen, O.; Wilting, P.; Koch, H.J.; Märländer, B. Yield and nitrogen use efficiency of fodder and sugar beet (*Beta vulgaris* L.) in contrasting environments of northwestern. Europe. *Eur. J. Agron.* **2016**, *73*, 124–132. [CrossRef]
10. Brauns, B.; Bjerg, P.L.; Song, X.F.; Jakobsen, R. Field scale interaction and nutrition exchange between surface water and shallow groundwater in the Baiyang Lake region, North China Plain. *J. Environ. Sci.* **2016**, *45*, 60–75. [CrossRef]
11. Yan, F.; Zhang, F.; Fan, X.; Fan, J.; Wang, Y.; Zou, H.; Wang, H.; Li, G. Determining irrigation amount and fertilization rate to simultaneously optimize grain yield, grain nitrogen accumulation and economic benefit of drip-fertigated spring maize in northwest China. *Agric. Water Manag.* **2020**, *243*, 106440. [CrossRef]
12. Kiymaz, S.; Ertek, A. Yield and quality of sugar beet (*Beta vulgaris* L.) at different water and nitrogen levels under the climatic conditions of Krsehir, Turkey. *Agric. Water Manag.* **2015**, *158*, 156–165. [CrossRef]
13. Olfs, H.-W.; Blankenau, K.; Brentrup, F.; Jasper, J.; Link, A.; Lammel, J. Soil- and plant-based nitrogen-fertilizer recommendations in arable farming. *J. Plant Nutr. Soil Sci.* **2005**, *168*, 414–431. [CrossRef]
14. Kunrath, T.R.; Lemaire, G.; Sadras, V.O.; Gastal, F. Water use efficiency in perennial forage species: Interactions between nitrogen nutrition and water deficit. *Field Crops Res.* **2018**, *222*, 1–11. [CrossRef]
15. Tognetti, R.; Palladino, M.; Minnocci, A.; Delfine, S.; Alvino, A. The response of sugar beet to drip and low-pressure sprinkler irrigation in southern Italy. *Agric. Water Manag.* **2003**, *60*, 135–155. [CrossRef]
16. Topak, R.; Acar, B.; Uyanöz, R.; Ceyhan, E. Performance of partial root-zone drip irrigation for sugar beet production in a semi-arid area. *Agric. Water Manag.* **2016**, *176*, 180–190. [CrossRef]
17. Zhou, H.; Wang, L.; Xu, P.; Liu, D.; Hao, Y.; Wang, K.; Fan, H. Silicon drip fertigation improved sugar beet root and canopy growth and alleviated water deficit stress in arid areas. *Eur. J. Agron.* **2024**, *152*, 127236. [CrossRef]
18. Vamerali, T.; Guarise, M.; Ganis, A.; Mosca, G. Effects of water and nitrogen management on fibrous root distribution and turnover in sugar beet. *Eur. J. Agron.* **2009**, *31*, 69–76. [CrossRef]
19. Tian, H.; Shi, S.; Wang, H.; Li, F.; Li, Z.; Alva, A.; Zhang, Z. Estimation of sugar beet aboveground biomass by band depth optimization of hyperspectral canopy reflectance. *J. Indian Soc. Remote* **2016**, *45*, 795–803. [CrossRef]
20. King, B.A.; Tarkalson, D.; Bjorneberg, D. Evaluation of canopy temperature based crop water stress index for deficit irrigation management of sugar beet in semi-arid climate. *Appl. Eng. Agric.* **2024**, *40*, 15822. [CrossRef]
21. Kiymaz, S.; Ertek, A. Water use and yield of sugar beet (*Beta vulgaris* L.) under drip irrigation at different water regimes. *Agric. Water Manag.* **2015**, *158*, 225–234. [CrossRef]
22. Li, Z. The Physiological Effects on Coupling of Water and Nitrogen in Sugar Beets (*Beta Vulgaris* L.) with Drip Irrigation Under Plastic Mulch. Doctoral Dissertation, Inner Mongolia Agricultural University, Hohhot, China, 2018. (In Chinese).
23. Bu, H.; Sharma, L.K.; Denton, A.; Franzen, D.W. Sugar beet yield and quality prediction at multiple harvest dates using active-optical sensors. *Agron. J.* **2016**, *108*, 273–284. [CrossRef]
24. Bu, H.; Sharma, L.K.; Denton, A.; Franzen, D.W. Comparison of satellite imagery and ground-based active optical sensors as yield predictors in sugar beet, spring wheat, corn, and sunflower. *Agron. J.* **2017**, *109*, 299–308. [CrossRef]
25. Greenwood, D.J.; Gastal, F.; Lemaire, G.; Draycott, A.; Millard, P.; Neeteson, J.J. Growth rate and %N of field grown crops: Theory and experiments. *Ann Bot.* **1991**, *67*, 181–190. [CrossRef]

26. Zhao, B.; Liu, Z.; Ata-UI-Karim, S.T.; Xiao, J.; Liu, Z.; Qi, A.; Ning, D.; Nan, J.; Duan, A. Rapid and nondestructive estimation of the nitrogen nutrition index in winter barley using chlorophyll measurements. *Field Crops Res.* **2016**, *185*, 59–68. [CrossRef]
27. Soratto, R.P.; Sandaña, P.; Sousa, W.S.; Fernandes, A.M.; Ciampitti, I. Critical nitrogen dilution curve for estimating nitrogen nutrition index of common beans. *Field Crops Res.* **2025**, *299*, 109713. [CrossRef]
28. Justes, E.; Mary, B.; Meynard, J.M.; Machet, J.M.; Thelierhuche, L. Determination of a critical nitrogen dilution curve for winter wheat crops. *Ann Bot.* **1994**, *74*, 397–407. [CrossRef]
29. Qiang, S.; Zhang, F.; Miles, D.; Zhang, Y.; Xiang, Y.; Fan, J. Determination of critical nitrogen dilution curve based on leaf area index for winter wheat in the Guanzhong Plain, Northwest China. *J. Integr. Agr.* **2019**, *18*, 2–13. [CrossRef]
30. Ravier, C.; Quemada, M.; Jeuffroy, M.H. Use of a chlorophyll meter to assess nitrogen nutrition index during the growth cycle in winter wheat. *Field Crops Res.* **2017**, *214*, 73–82. [CrossRef]
31. Greenwood, D.J.; Lemaire, G.; Gosse, G.; Cruz, P.; Draycott, A.; Neeteson, J.J. Decline in percentage N of C3 and C4 crops with increasing plant mass. *Ann Bot.* **1990**, *66*, 425–436. [CrossRef]
32. Lemaire, G.; Gastal, F. N uptake and distribution in plant canopies. In *Diagnosis of the Nitrogen Status in Crops*; Lemaire, G., Ed.; Springer: Berlin/Heidelberg, Germany, 1997; pp. 3–43.
33. Lu, J.; Hu, T.; Li, Y.; Cui, X.; Cheng, M.; Yan, S.; Xiang, Y.; Fan, J.; Zhang, F.; Xiong, Y. Effects of different water management strategies on critical nitrogen concentration dilution curves, nitrogen accumulation, and grain yield in winter wheat. *Agric. Commun.* **2024**, *2*, 100052. [CrossRef]
34. Li, W.; He, P.; Jin, J. Critical nitrogen curve and nitrogen nutrition index for spring maize in north-east China. *J. Plant Nutr.* **2012**, *35*, 1747–1761. [CrossRef]
35. Yue, S.; Sun, F.; Meng, Q.; Zhao, R.; Li, F.; Chen, X.; Zhang, F.; Cui, Z. Validation of a critical nitrogen curve for summer maize in the North China Plain. *Pedosphere* **2014**, *24*, 76–83. [CrossRef]
36. Yao, X.; Ata-Ul-Karim, S.T.; Zhu, Y.; Tian, Y.; Liu, X.; Cao, W. Development of critical nitrogen dilution curve in rice based on leaf dry matter. *Eur. J. Agron.* **2014**, *55*, 20–28. [CrossRef]
37. Giletto, C.M.; Echeverría, H.E. Critical nitrogen dilution curve for processing potato in Argentinean humid pampas. *Am. J. Potato Res.* **2012**, *89*, 102–110. [CrossRef]
38. Tei, F.; Benincasa, P.; Guiducci, M. Critical nitrogen concentration in processing tomato. *Eur. J. Agron.* **2002**, *18*, 45–55. [CrossRef]
39. Yan, F.; Zhang, F.; Fan, J.; Hou, X.; Bai, W.; Liu, X.; Wang, Y.; Pan, X. Optimization of irrigation and nitrogen fertilization increases ash salt accumulation and ions absorption of drip-fertigated sugar beet in saline-alkali soils. *Field Crops Res.* **2021**, *271*, 108247. [CrossRef]
40. Allen, R.G.; Pereira, L.S.; Raes, D.; Smith, M. *Crop Evapotranspiration: Guidelines for Computing Crop Water Requirements*; Food and Agriculture Organization of the United Nations: Rome, Italy, 1998.
41. Li, Y.; Zhou, M. Projections of water requirements of cotton and sugar beet in Xinjiang based on statistical downscaling model. Trans. *Chin. Soc. Agric. Eng.* **2014**, *30*, 70–79.
42. Mahmoud, E.A.; Hassanin, M.A.; Borham, T.I.; Emara, E.I.R. Tolerance of some sugar beet varieties to water stress. *Agric. Water Manag.* **2018**, *201*, 144–151. [CrossRef]
43. Ebmeyer, H.; Hoffmann, C.M. Efficiency of nitrogen uptake and utilization in sugar beet genotypes. *Field Crops Res.* **2021**, *274*, 108334. [CrossRef]
44. Chilundo, M.; Joel, A.; Wesström, I.; Brito, R.; Messing, I. Influence of irrigation and fertilisation management on the seasonal distribution of water and nitrogen in a semi-arid loamy sandy soil. *Agric. Water Manag.* **2018**, *199*, 120–137. [CrossRef]
45. Gastal, F.; Lemaire, G.; Durand, J.L.; Louarn, G. Chapter 8—Quantifying crop responses to nitrogen and avenues to improve nitrogen-use efficiency. In *Crop Physiology*; Academic Press: Cambridge, MA, USA, 2015; pp. 161–206.
46. Fitzgerald, G.; Rodriguez, D.; Leary, G. Measuring and predicting canopy nitrogen nutrition in wheat using a spectral index-The canopy chlorophyll content index (CCCI). *Field Crops Res.* **2010**, *116*, 318–324. [CrossRef]
47. Ata-Ul-Karim, S.T.; Zhu, Y.; Yao, X.; Cao, W.X. Determination of critical nitrogen dilution curve based on leaf area index in rice. *Field Crops Res.* **2014**, *167*, 76–85. [CrossRef]
48. Lemaire, G.; Jeuffroy, M.H.; Grancois, F. Diagnosis tool for plant and crop Nstatus in vegetative stage: Theory and practices for crop N management. *Eur. J. Agron.* **2008**, *28*, 614–624. [CrossRef]
49. Weeden, B.R. *Potential of Sugar Beet on the Atherton Tableland: A Report for the Rural Industries Research and Development Corporation*; RIRDC: Wagga Wagga, Australia, 2000.
50. Ghimire, D.; Maharjan, B. Optimizing nitrogen management to enhance irrigated sugar beet yield and quality. *Agron. J.* **2024**, *116*, 2564–2572. [CrossRef]
51. Armstrong, M.J.; Milford, G.F.J.; Pocock, T.O.; Last, P.J.; Day, W. The dynamics of nitrogen uptake and its remobilization during the growth of sugar beet. *J. Agric. Sci.* **1986**, *107*, 145–154. [CrossRef]

52. Quiñones, A.; Martínez-Alcántara, B.; Legaz, F. Influence of irrigation system and fertilization management on seasonal distribution of N in the soil profile and on N-uptake by citrus trees. *Agric. Ecosyst. Environ.* **2007**, *122*, 399–409. [CrossRef]
53. Rathke, G.; Behrens, T.; Diepenbrock, W. Integrated nitrogen management strategies to improve seed yield, oil content and nitrogen efficiency of winter oilseed rape (*Brassica napus* L.): A review. *Agric. Ecosyst. Environ.* **2006**, *117*, 80–108. [CrossRef]
54. Gorska, A.; Qing, Y.; Holbrook, N.M.; Zwieniecki, M.A. Nitrate control of root hydraulic properties in Plants: Translating local information to whole plant response. *Plant Physiol.* **2008**, *148*, 1159–1167. [CrossRef]
55. Guo, J.; Fan, J.; Zhang, F.; Yan, S.; Zheng, J.; Wu, Y.; Li, J.; Wang, Y.; Sun, X.; Liu, X.; et al. Blending urea and slow-release nitrogen fertilizer increases dryland maize yield and nitrogen use efficiency while mitigating ammonia volatilization. *Sci. Total Environ.* **2021**, *790*, 148058. [CrossRef]
56. Singh, A.; Rudnick, D.; Snow, D.; Misar, C.; Birru, G.; Proctor, C.; Puntel, L.; Iqbal, J. Intra-and inter-annual variability of nitrogen and irrigation management effects on nitrate leaching and maize yield in the Bazile Groundwater Management Area, Nebraska. *Agric. Ecosyst. Environ.* **2025**, *381*, 109463. [CrossRef]
57. Badr, M.A.; Hussein, S.D.A.; Eltohamy, W.A.; Gruda, N. Nutrient uptake and yield of tomato under various methods of fertilizer application and levels of fertigation in arid lands. *Gesunde Pflanzen* **2010**, *62*, 11–19. [CrossRef]
58. Ziadi, N.; Bélanger, G.; Claessens, A.; Lefebvre, L.; Cambouris, A.N.; Tremblay, N.; Parent, L.É. Determination of a critical nitrogen dilution curve for spring wheat. *Agron. J.* **2010**, *102*, 241–250. [CrossRef]
59. Ziadi, N.; Brassard, M.; Bélanger, G.; Cambouris, A.N.; Tremblay, N.; Nolin, M.C.; Claessens, A.; Parent, L.É. Critical nitrogen curve and nitrogen nutrition index for corn in eastern Canada. *Agron. J.* **2008**, *100*, 271–276. [CrossRef]
60. Fabbri, C.; Mancini, M.; Marta, A.D.; Orlandini, S.; Napoli, M. Integrating satellite data with a nitrogen nutrition curve for precision top-dress fertilization of durum wheat. *Eur. J. Agron.* **2020**, *120*, 126148. [CrossRef]
61. Malnou, C.S.; Jaggard, K.W.; Sparkes, D.L. Nitrogen fertilizer and the efficiency of the sugar beet crop in late summer. *Eur. J. Agron.* **2008**, *28*, 47–56. [CrossRef]
62. Qiang, S.; Zhang, F.; Xiang, Y.; Zhang, Y.; Yan, S.; Xing, Y. Simulation and validation of critical nitrogen dilution curve for summer maize in Guanzhong Plain during different rainfall years. *Trans. Chin. Soc. Agric. Eng.* **2015**, *31*, 168–175.
63. Cheng, M.; He, J.; Wang, H.; Fan, J.; Xiang, Y.; Liu, X.; Liao, Z.; Tang, Z.; Abdelghany, A.E.; Zhang, F. Establishing critical nitrogen dilution curves based on leaf area index and aboveground biomass for greenhouse cherry tomato: A Bayesian analysis. *Eur. J. Agron.* **2022**, *141*, 126615. [CrossRef]

MDPI AG
Grosspeteranlage 5
4052 Basel
Switzerland
Tel.: +41 61 683 77 34

Plants Editorial Office
E-mail: plants@mdpi.com
www.mdpi.com/journal/plants

www.ingramcontent.com/pod-product-compliance
Lightning Source LLC
Chambersburg PA
CBHW062034220926
51850CB00010B/2538

* 9 7 8 3 7 2 5 8 7 3 2 8 9 *